Design and Implementation of Soft Robotic Manipulators

Xiang Zhang · Kangjia Fu · Xuesong Wu ·
Hongwei Liu

Design and Implementation of Soft Robotic Manipulators

Science Press
Beijing

Springer

Xiang Zhang
Defense Innovation Institute
Academy of Military Sciences
Beijing, China

Kangjia Fu
Defense Innovation Institute
Academy of Military Sciences
Beijing, China

Xuesong Wu
Defense Innovation Institute
Academy of Military Sciences
Beijing, China

Hongwei Liu
Defense Innovation Institute
Academy of Military Sciences
Beijing, China

ISBN 978-981-95-1656-8 ISBN 978-981-95-1657-5 (eBook)
https://doi.org/10.1007/978-981-95-1657-5

Jointly published with Science Press, Beijing, China
The print edition is not for sale in China Mainland. Customers from China Mainland please order the print book from: Science Press.
ISBN of Co-Publisher's edition: 978-7-03-083693-9

This Springer imprint is published by the registered company Springer Nature Singapore Pte Ltd.
The registered company address is: 152 Beach Road, #21-01/04 Gateway East, Singapore 189721, Singapore

Foreword

Soft manipulator is a kind of special manipulator that uses soft material or flexible structure to perform manipulation tasks under the specific drive mode inspired by the soft tissue. It is developing gradually as an important component of the soft robot. The wide range of abilities demonstrated by soft tissues, such as elephant trunks and octopus tentacles, have motivated a recent surge of research activity in soft manipulators. Compared with conventional rigid manipulators, soft manipulators are inherently flexible and compliant, enabling them to easily change their deformation for a given task. These advantages enable soft manipulators to achieve safe interaction and flexible manipulation in complex and narrow environments such as search and rescue operations in collapsed buildings, underwater grabbing and medical operations, gradually bringing about thriving development in lots of robotic applications.

The realization of soft manipulators tends to be interesting but challenging. I am very familiar with the authors of this monograph because they were all my Ph.D. students. They have been working on the soft robotic manipulator for almost ten years. I still remember clearly the difficult but happy time when they first started working on this topic. They have investigated hundreds of typical design cases and developed several kinds of novel soft manipulators. In particular, there are not only the design and construction of the soft manipulator hardware but also modelling and application algorithms like kinematic modelling and tensile damage analysis. Now the authors published their achievements, aiming to provide theoretical guidance and technical support for researchers working on robotics, especially soft manipulators.

This monograph aims to investigate the design and implementation of soft robotic manipulators. It starts with a review of soft manipulator research, applications and opportunities and then provides a comprehensive presentation from design and simulation to experiment validations of soft manipulators, revealing deep insight into fine control and suitable application. This monograph has the following three characteristics. Firstly, the research advance and prospect of the soft manipulator worldwide are investigated. The researchers in related fields can have a quick and comprehensive understanding of the soft robotic manipulators by referring this book. Secondly, the authors give a detailed and thorough account of the team's recent ten years of research work on this topic, including the design and development of an elephant trunk-like soft robotic manipulator, a foldable pneumatic soft manipulator and a cable-driven continuum manipulator. Thirdly, the authors have carried out a lot of theoretical modelling and motion control research for different types of soft manipulators and formed very valuable theoretical research results, which provides a good theoretical reference for the subsequent research of soft robot.

I'm glad to see this monograph get published. I believe that it will provide a useful reference for solving related technical problems for graduate students, researchers and engineers who are interested in bio-robot, mechanism design, intelligent control, etc.

Xiaoqian Chen
Academician
Chinese Academy of Sciences
Beijing, China

Preface

Inspired by the movement of soft tissues such as the elephant trunk and octopus's tentacles, the soft manipulator is a kind of special manipulator that uses soft material or flexible structure to perform manipulation tasks through continuous motion control under the action of specific continuous drive mode. Due to its advantages of low weight, flexibility and safe physical interaction, the soft robotic manipulator shows a wide prospect in both industry and family applications. As an important component of the soft robot, the soft manipulator is developing step by step with the fast development of the soft robot. In the Web of Science database, the search terms or search scopes including "soft manipulator", "soft robot arm" and "soft arm" show a booming trend. According to different drive modes, the soft manipulator can be divided into fluid-driven mode, cable-driven mode, smart material-driven mode, electromagnetic-driven mode, and so on. Among them, the smart material-driven mode includes the electroactive polymer-driven mode and the shape memory material-driven mode. Moreover, the hybrid of two different drive modes is also studied in order to improve the control accuracy or the stability of the soft manipulator. For different application fields, the selection of the design scheme and drive mode of the soft manipulator needs to be analyzed according to the specific requirements.

Currently, the research on the soft manipulator has initially formed a technical system, although it is still in the primary stage. We have summarized the enabling technologies systematically, such as the integrated design of structure and actuation, kinematic and dynamic modeling, motion control, shape detection, dynamic tactile perception and stiffening method. With the development of the soft manipulator, its overall technology maturity will be gradually improved. The soft robotic manipulator, which has advantages of low weight, flexibility and safe physical interaction, shows a wide prospect in both industry and family applications. Currently, the applications of the soft manipulator lie mainly in the fields of the ground rescue, underwater grabbing, medical operation, assist maintenance, man-machine interaction, space manipulation and so on. The application fields, such as the origami soft manipulator, biomimetic soft microrobots, self-healing soft robot, space soft manipulator and so on, emerge new opportunities and challenges for the soft manipulators.

Our research group has focused on the soft robotic manipulator for nearly ten years. We investigated hundreds of the typical design cases of the soft manipulators as well as the related research teams. Moreover, we design and develop several kinds of soft manipulators, which deepen our understanding of this field. First, we design a pneumatic soft manipulator, which is made up of three pneumatic soft actuators stacked in series. The soft actuator is composed of three inflatable bellows, which are connected and constrained by a series of thin rigid plates. The bellows is stretched by the actuation of pressured gas. The cooperation of the bellows and the rigid constraints affect the soft actuator's directional movement. The stacking connection of several soft actuators can have a large coverage in space. Then, a structural analysis of a pneumatic soft manipulator is performed, and an optimization design method for the inclination angle and supporting reaction force is proposed. Next, the reliability of each actuator is evaluated in order to make sure the safety of the soft manipulator. The reliability of the composite pneumatic actuator is studied by theoretical modeling, FEM simulation and experiment. An improved design is carried out for the scheme of trunk-like soft manipulator, so as to make it conform to the variable cross-section configuration of elephant trunk biological tissue. The corresponding path planning and trajectory control are carried out based on an improved prototype and test system of the trunk-like soft manipulator. Moreover, a foldable pneumatic soft manipulator is proposed in our investigation. The soft manipulator comprises three pneumatic actuators and foldable straight arms and can transfer its shape between the contracted and expanded states, making its volume foldable. According to this design, a relevant kinematic and inverse kinematic model for the soft manipulator is proposed. The shape deformation and workspace are numerically calculated and analyzed. In the final part of this book, a cable-driven continuum manipulator as well as its soft gripper is designed and developed. The continuum manipulator is made up of two sections, including nine threading plates and eight spring steels. The grasping force and the wrapping area are both considered into the topological optimization process of the soft gripper in order to get a satisfying grasping performance. The grasping performance of the soft gripper and the motion control of the continuum arm are validated experimentally.

We hereby would like to sincerely thank the National Natural Science Foundation of China (12472351, 12102477, 12402411, and 62303487) for the farsighted support. We would especially thank academician Xiaoqian Chen for contributing forward of this book, as well as his consistent care and support for our research. We thank Ms. Jie Chen, the editor, for her great efforts in getting this book published. In the process of writing this academic monograph, we read a large amount of literature and cited some of the most relevant literature in the book. Here, we would like to express our thanks.

The soft robotic manipulator is an emerging frontier crossing field, including mechanics, machinery, control, materials, electronics, information, etc. Moreover, it is also one of the most ideal application platforms for embodied artificial intelligence. However, its theoretical exploration and practical application are still in the initial stage, and there are still many key technologies that need to be further studied. This book is a systematic summary of the relevant research work of the authors' team in recent years. Due to the limited level of the author, it is difficult to avoid omissions. Finally, we sincerely welcome any corrections from peers and readers for errors or potential points of confusion in this book.

Xiang Zhang
Beijing, China

Contents

Chapter 1
Review of Soft Manipulator Research, Applications and Opportunities

Dexterous manipulation is one of the primary goals in robotics [1]. Inspired by the movement of soft tissues such as the elephant trunk and octopus's tentacles, the soft manipulator is a kind of special manipulator that uses soft materials or flexible structure to perform manipulation tasks through continuous motion control under the action of specific continuous drive mode. The definition of soft material is based on the elastic modulus of the biological tendon (about 1 GPa), which can be regarded as soft material if it is less than 1 GPa [2]. It is an under-actuated system that performs infinite freedom motion through limited control variables [3, 4], which is essentially different from the rigid manipulator [5]. The control variables of soft manipulators are closely related to their drive modes. At present, the main drive modes include fluid drive, cable drive, smart material drive, electromagnetic drive and so on [2]. The fluid drive is the use of air pressure or hydraulic pressure to drive the deformation of the capsule. The cable drive is the use of motors to control the cables. The smart material drive is actuated by the deformation of smart materials or structures through the change of voltage or current [6]. The electromagnetic drive is the use of the external controllable magnetic field to drive the ferromagnet.

As an important component of the soft robot, the soft manipulator is developing step by step with the fast development of the soft robot. According to different drive modes, the relevant-typical design cases of the soft manipulators are investigated. In terms of fluid-driven mode, Wehner et al. [7] designed and developed an entirely soft robot Octobot by using a multi-material, embedded 3D printing technique. Sanan et al. [8, 9] designed and developed an arm-like soft manipulator PneuArm, which was composed of multiple air bags stacked and assembled in a complex way, without any rigid skeleton structure. Chitrakaran et al. [10, 11] designed and developed a soft manipulator OctArm imitating the tentacle of the octopus. In terms of the cable-driven mode, Dong et al. [12–14] developed a slender continuous soft manipulator, which was mainly used for nondestructive testing and the inspection of aeroengine. Li et al. [15] designed and developed a cable-driven continuous backbone robot, which was composed of a bendable backbone and a series of circular plates with holes. Giorelli et al. [16, 17] designed and developed a soft manipulator imitating

X. Zhang et al., *Design and Implementation of Soft Robotic Manipulators*,
https://doi.org/10.1007/978-981-95-1657-5_1

octopus tentacle, in which 12 strings were embedded into the soft silica gel. In terms of smart material-driven mode, Kovacs et al. [18] designed and developed a stacked dielectric elastomer actuator, which was made by brushing graphite electrode on the dielectric elastomer film and then stacking the film into a cylindrical actuator. Pei et al. [19, 20] designed and developed a multiple-degrees-of-freedom electroelastomer roll actuator by rolling highly prestrained electroelastomer films around a compression spring. Cianchetti et al. [21, 22] designed and developed an octopus like soft manipulator inspired by the biological mechanism of octopus' tentacle. In terms of electromagnetic-driven mode, Kim et al. [23] designed and developed a ferromagnetic soft continuum robot, which was mainly used for minimally invasive cleaning of blood vessels, based on their existing technologies such as programmable magnetic printing [24] and hydrogel bioelectronics devices [25]. Jeon et al. [26, 27] designed and developed a magnetically controlled soft microrobot, which was supported by a complex external electromagnetic mechanism. All these above investigations indicate that the researches of soft manipulators have attracted worldwide interest and far-reaching influence.

In this chapter, a review of research progresses, applications and opportunities of the soft manipulators is summarized and analyzed through plentiful literature investigations. Firstly, the typical design cases of the soft manipulators are systematically classified according to the drive modes. The performances of different drive modes are compared and analyzed. Then, the enabling technologies involved in the research of the soft manipulator are discussed from the aspects of the modeling and control, the shape detection and dynamic tactile sensing, the stiffening method, and so on. Finally, the current application fields of the soft manipulator as well as its opportunities and challenges in the newly booming fields are analyzed and described.

1.1 Related Literature Analysis

In the Web of Science database, "soft manipulator", "soft robot arm" and "soft arm" are used as the search terms and the search scope is "theme", covering the period from 2000 to April 2021. Irrelevant research directions were excluded, with a total of 4970 records detected. Figure 1.1 shows the current research subjects and research hotspots of soft manipulators. Citespace software was selected as a graphical analysis tool to analyze the data. The mapping knowledge domain of the soft manipulator is shown in Fig. 1.1.

As can be seen in Fig. 1.1a, USA is still leading the research of soft manipulators worldwide, followed by China, Germany, England, Italy and Japan, etc. The most famous institutes in USA include Harvard University, Johns Hopkins University, Stanford University, University of Michigan, etc. The famous institutes in China include China Academy Science, Shanghai Jiao Tong University, Harbin Institute of Technology, etc., as is shown in Fig. 1.1b. The cluster analysis of categories in Fig. 1.1c shows that soft manipulators belong to an interdisciplinary field and involve many research fields, such as engineering, surgery, orthopedics, oncology,

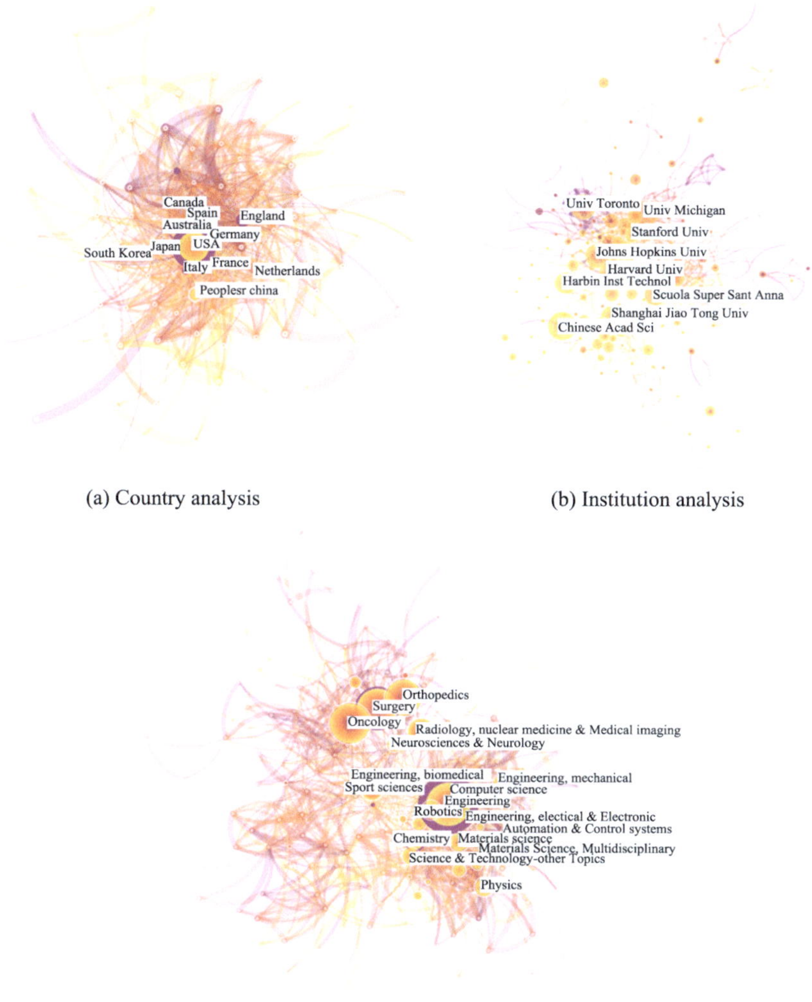

(a) Country analysis

(b) Institution analysis

(c) Cluster analysis of categories

Fig. 1.1 The mapping knowledge domain of soft manipulator

robotics, materials science, chemistry, electrical and electronic, physics, and so on. The number of papers that varies with the year is shown in Fig. 1.2. It can be seen that the number of papers booms up in the year of 2002, and then increases steadily. This suggests that the researchers have begun to focus on soft manipulators from 2002, and the research is getting more and more attention and focus steadily. From the mapping knowledge analysis above, it is concluded that the research of the soft manipulator belongs to an interdisciplinary research field and has already become a hot topic worldwide.

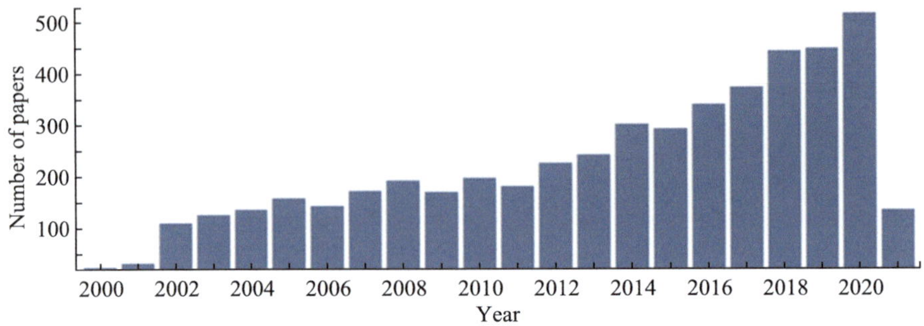

Fig. 1.2 The number of papers focusing on the soft manipulators varies with the year

1.2 Research Status of Soft Manipulator

According to different drive modes, the soft manipulator can be divided into fluid-driven mode, cable-driven mode, smart material-driven mode, electromagnetic-driven mode, and so on. Among them, the smart material-driven mode includes the electroactive polymer driven mode and the shape memory material driven mode. Moreover, the hybrid of two different drive modes is also studied in order to improve the control accuracy or the stability of the soft manipulator. In the following, the typical design cases of the soft manipulators related to different drive modes are introduced respectively.

1.2.1 Fluid-Driven Mode

The soft manipulator with the fluid-driven mode makes spatial movement by controlling the gas or liquid pressure in the multi-cavity structure. According to this definition, the fluid-driven mode can be divided into gas-driven mode and liquid-driven mode. The gas-driven mode is light and pollution-free, which is widely used at present; the liquid-driven mode has a high response frequency due to the incompressibility of the liquid [28]. The difficulty of the fluid-driven mode lies in the design and the pressure control of the multi-cavity structure. Currently, a lot of research works about this drive mode have been done and reported in the literature. The typical design cases are shown in Fig. 1.3.

Sanan et al. [8, 9, 52, 53] designed and developed an arm-like soft manipulator PneuArm, which was composed of multiple air bags stacked and assembled in a complex way, without any rigid skeleton structure, as shown in Fig. 1.3a. Walker et al. [10, 11, 54–59] designed and developed a soft manipulator OctArm, which imitated the tentacle of the octopus. It was composed of multiple soft actuators stacked in series, and each actuator adopted a symmetrical parallel structure with three cavities, as shown in Fig. 1.3b. OctArm had developed to the sixth generation. Its overall

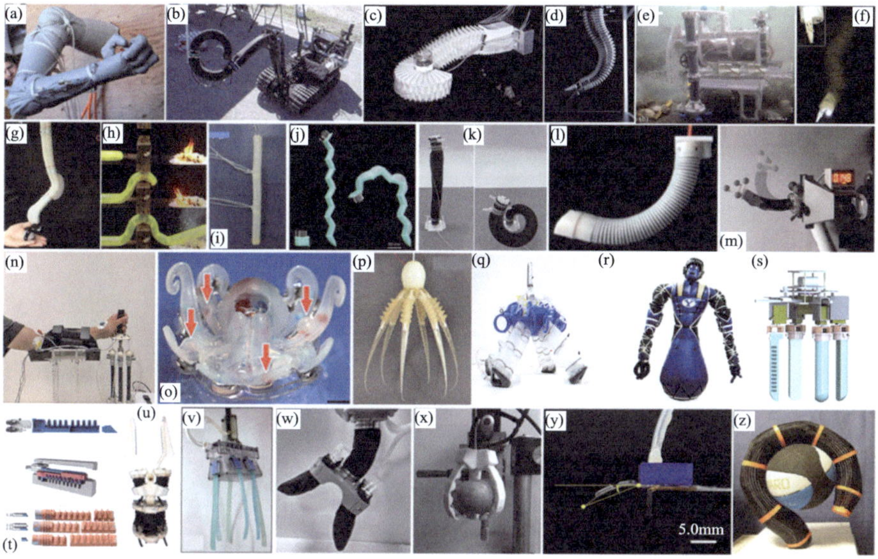

Fig. 1.3 Fluid-driven soft manipulator: **a** PneuArm [9]; **b** OctArm [10]; **c** HPN [29]; **d** BHA [30]; **e** underwater soft manipulator [31]; **f** multi-module [32]; **g** fluidic elastomer [33]; **h** pressure-driven lengthening [34]; **i** omnidirectional bending [35]; **j** soft robots consisting of dual-morphing M-ori [36]; **k** helical contractile PAM (HC-PAM) [37]; **l** flexible hybrid pneumatic actuator (FHPA) [38]; **m** Spring reinforced actuator (SRA) [39]; **n** a soft parallel robot [40]; **o** Octobot [7]; **p** soft hydraulic robot [41]; **q** soft-legged untethered quadruped robot [42]; **r** inflatable humanoid robot named King Louie [43]; **s** a variable structure pneumatic soft robot [44]; **t** multi-segment soft robotic fingers [45]; **u** Underwater robotic manipulator [46]; **v** nanofiber-reinforced soft robotic actuators [47]; **w** fabric-based versatile and stiffness-tunable soft gripper [48]; **x** three-finger soft pneumatic gripper [49]; **y** soft pneumatic microactuators [50]; **z** robotic continuum manipulator driven by PMA [51]

handling performance was gradually improved by enhancing the bearing capacity of the structure, adding the functions of shape detection and rotatable base. Chen et al. [29, 60–62] designed a soft manipulator based on Honeycomb Pneumatic Networks (HPN), as shown in Fig. 1.3c. The soft manipulator was demonstrated to have a stable control performance due to the structural support of the honeycomb frame. Hildebrandt et al. [30, 63–66] designed and developed a Bionic Handling Assistant (BHA), which was composed of a soft manipulator and a soft gripper, as shown in Fig. 1.3d. The soft manipulator was composed of three pneumatic actuators, which were composed of pneumatic bellows connected in parallel by a series of thin plates. Gong et al. [31] designed and developed a three-degree-of-freedom (3-DOF) soft manipulator with an opposite-bending-and-stretching structure (OBSS) for spatial delicate grasping in shallow water, as shown in Fig. 1.3e. Cianchetti et al. [32, 67–71] designed and developed a soft multi-module manipulator for medical minimally invasive surgery, as shown in Fig. 1.3f. It was composed of three pneumatic actuators, which were all parallel structures of three air chambers. The motion control of the actuator was realized through the independent air pressure change of the air chambers. Rus et al. [33,

72–76] designed and developed a soft fluidic elastomer manipulator, as shown in Fig. 1.3g. The team started with a planar actuator consisting of two symmetrical parallel air chambers [73, 74], and then the planar actuator was extended to a spatial actuator by arranging four cavities in a symmetrical parallel way [72, 75]. Hawkes et al. [34, 77–81] designed and developed a soft manipulator that could grow automatically to adapt the change of the surrounding environment, which was inspired by the biological principles of neurons, pollen tubes and thick-walled tissue cells, as shown in Fig. 1.3h. Yan et al. [35, 82, 83] proposed a pneumatic omnidirectional bending soft manipulator, which was composed of multi-stage pneumatic omnidirectional soft actuators in series, as shown in Fig. 1.3i. The soft actuator was composed of three independent parallel air chambers formed symmetrically in parallel. Kim et al. [36] introduced pelican eel-inspired dual-morphing architectures that embodied quasi-sequential behaviors of origami unfolding and skin stretching in response to fluid pressure, as shown in Fig. 1.3j. Guan et al. [37] proposed and analyzed a family of bending and helical extensile PAM (HE-PAM)/helical contractile PAM (HC-PAM) under the inspiration of the elephant trunk, as shown in Fig. 1.3k. Liu et al. [38] presented a soft humanoid hand that was made of flexible hybrid pneumatic actuators (FHPA) designed based on a modular approach, as shown in Fig. 1.3l. Fu et al. [39] proposed a spring reinforced actuator (SRA) that explored the intermediate state between muscular hydrostats and endo-skeletal mechanisms, as shown in Fig. 1.3m. Wang et al. [40] designed and developed a soft parallel robot based on pneumatic artificial muscles for the wrist rehabilitation, as shown in Fig. 1.3n. Wehner et al. [7] designed and developed an entirely soft robot Octobot by using a multi-material, embedded 3D printing technique, as shown in Fig. 1.3o. The soft robot adopted the integrated design and fabrication technology of multi soft materials, and its internal pressure change was provided by the release of gas through the chemical reaction of H_2O_2 solution under the action of platinum catalyst [84, 85]. Zatopa et al. [41] presented an octopus-like robot with completely soft hydraulic control valves consisting of a 3D-printed photopolymer body with electrorheological (ER) fluid as a working fluid and gallium-indium-tin liquid metal alloy as electrodes, as shown in Fig. 1.3p. Drotman et al. [42] proposed a soft-legged untethered quadruped robot with a bioinspired gait pattern controlled with an electronics-free pneumatic actuation system, as shown in Fig. 1.3q. Killpack et al. [43] designed and developed an inflatable humanoid robot named King Louie, who had two pneumatic soft arms, as shown in Fig. 1.3r. Huang et al. [44] proposed a variable structure pneumatic soft robot, whose structure was variable in that when it grasped irregular objects, it could adapt to different sizes by active expansion or contraction, as shown in Fig. 1.3s. Teeple et al. [45] designed soft robotic fingers for robust precision grasping, as shown in Fig. 1.3t. They found that antipodal grasps were more stable when contacted with the object occurred on the side of the fingers instead of the fingertips. Shen et al. [46] developed an underwater robotic manipulator with a hybrid structure (rigid frame coupled with soft actuators) for improved rigidity and force output. The uniform design allowed one compact hydraulic actuation system to drive all actuators, as shown in Fig. 1.3u. Sinatra et al. [47] presented a nanofiber-reinforced soft robotic actuators capable of grasping delicate specimens of gelatinous marine life, as shown in Fig. 1.3v. Fei et al.

[48] presented a design of a fabric-based versatile and stiffness-tunable soft gripper integrating soft pneumatic fingers and wrist, which were versatile and adaptable to gripping objects of various shapes, sizes, and orientations in a wide range, as shown in Fig. 1.3w. Tawk et al. [49] demonstrated a three-finger soft pneumatic gripper made up of 3D printable soft vacuum actuators that were inspired by the sporangium of fern trees, as shown in Fig. 1.3x. Konishi et al. [50] presented a soft microactuator (a pneumatic balloon actuator) integrated with a flexible temperature sensor for a functional microfinger, as shown in Fig. 1.3y. Kang et al. [51] presented a robotic continuum manipulator driven by pneumatic muscle actuators (PMA) that was inspired by biological continuum structures, as shown in Fig. 1.3z. Currently, the soft manipulator was designed mostly by intuition, which complicated design improvement and hampered its advancement. Facing these challenges, Garriga-Casanovas et al. [86] presented a study of the design of soft manipulators with fluidic actuation using a common framework.

1.2.2 Cable-Driven Mode

The soft manipulator with the cable-driven mode makes the spatial motion by controlling the multi-cable system. Each cable needs a motor for its retraction and release control. It is necessary to configure the corresponding preloading device because the cable can only provide tensile force. In the research of the cable-driven mode, a large number of prototypes have been seen in the literature, as shown in Fig. 1.4.

Dong et al. [12–14, 94–97] designed and developed a slender continuous soft manipulator, which was mainly used for nondestructive testing and inspection of aeroengine. Its curl and extension states were shown in Fig. 1.4a. The soft manipulator had more than 25 control variables and a total length of 1.2 m that met the reach range of enough depth inside the engine. Rahn et al. [15] designed and developed a cable-driven continuous backbone robot, in which the central backbone was made of spring steel with a rectangular cross-section. A series of round plates with holes were installed along the central backbone at a certain interval, and the bending motion of the soft robot was controlled by controlling the cables, as shown in Fig. 1.4b. Laschi et al. designed and developed a soft manipulator imitating the octopus tentacle, which was composed of three soft actuations driven by 12 cables, as shown in Fig. 1.4c [16, 17, 98–101]. Voisembert et al. [87, 102–106] developed a long-range inflatable robotic arm, which was in fact a whole air bad with several bending joints formed by stop-pins or loosen wires, as shown in Fig. 1.4d. Liang et al. designed a multi-section tendon-driven continuous manipulator, which was composed of a series of rigid sections and omnidirectional joints, as shown in Fig. 1.4e [88, 107–109]. It had a total length of about 3 m and an external diameter of 0.35 m. Wang et al. [89, 110–115] presented a soft manipulator imitating the octopus tentacle and carried out a lot of research on the modeling, control and application of the soft manipulator, as shown in Fig. 1.4f. They developed a shape detection mechanism using fiber Bragg brating (FBG) to provide real-time feedback of the shape deformation for the

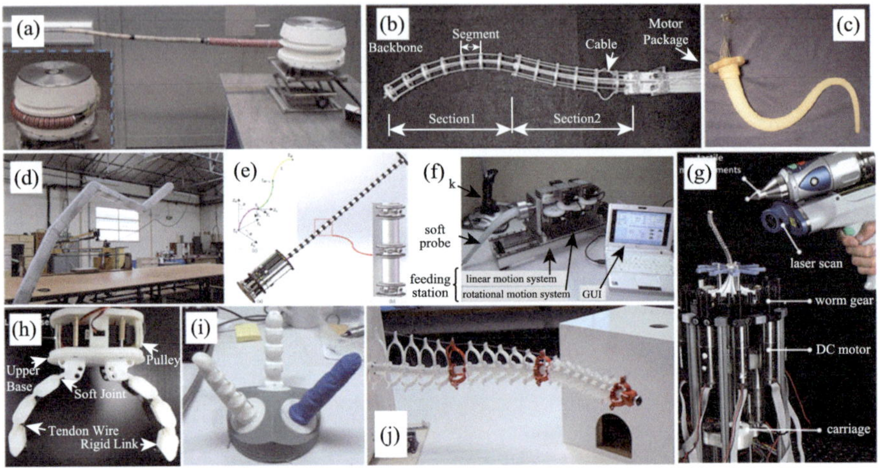

Fig. 1.4 Cable-driven soft manipulator: **a** slender continuum robot [14]; **b** continuous backbone [15]; **c** multibending soft robot [17]; **d** long-range inflatable arm [87]; **e** multi-section tendon-driven continuous manipulator [88]; **f** soft probe [89]; **g** tendon-driven continuum robot [90]; **h** soft-rigid tendon-driven modular grippers [91]; **i** soft monolithic fingers [92]; **j** the Echelon 3 deformable manipulator [93]

closed-loop motion control of the soft manipulator. Amanov et al. [90] presented a three-section tendon-driven continuum robot prototype with a diameter of 7 mm, which might be used in the scenarios of minimally invasive surgery and technical inspections, as shown in Fig. 1.4g. Hussain et al. [91, 116] developed soft-rigid tendon-driven modular grippers using interpenetrating phase composites materials, as shown in Fig. 1.4h. Mutlu et al. [92] designed a soft and monolithic robotic finger, which was controlled by using just one tendon and fabricated by 3D printing technique, as shown in Fig. 1.4i. Morales et al. [93] presented a dexterous, deformable manipulator bioinspired by the skeletal spine found in vertebrate animals, as shown in Fig. 1.4j.

1.2.3 Electroactive Polymer-Driven Mode

The soft manipulator with the electroactive polymer (EAP) driven mode makes spatial motion by the electrostatic attraction of the conductive layer to drive the multilayer film compress and deform. Electroactive polymers can be divided into electronic type and ionic type. The dielectric elastomer belongs to a typical electronic type. It requires a high drive voltage and can produce a relatively large force. The ionic type requires a low voltage, but it depends on the liquid environment and produces a small force. The typical design cases seen in the literature are investigated and shown in Fig. 1.5.

Gavrilovich et al. [117, 125–127] designed and developed a dielectric elastomer minimum energy structure (DEMES), which was mainly designed for the space debris removal task, as shown in Fig. 1.5a. Pei et al. [19, 20, 128, 129] presented a walking robot formed by assembling six actuators, which were composed of a central compression spring and a dielectric elastomer stacked structure, as shown in Fig. 1.5b. Choi et al. [118, 119, 130, 131] designed and developed an antagonistically driven linear actuator (ANTLA), which was stacked by multi-components of the dielectric elastomer in series, as shown in Fig. 1.5c. Inspired by the biological mechanism of the earthworm, Jang et al. [119] developed a multilayer crawling soft robot by distributing six multilayer dielectric elastomer actuators uniformly on the PCB board, as shown in Fig. 1.5d. Kovacs et al. [18, 132, 133] presented an active structure driven by stacked dielectric electro-active polymers (EAP), as shown in Fig. 1.5e. The active structure was composed of an array of equal segments, where the dielectric films were arranged in a pile-up configuration. Lipson et al. [120] presented a soft artificial muscle, which was composed of ethanol distributed throughout the solid silicone elastomer matrix, as shown in Fig. 1.5f. Zhao et al. [121] reported a somatosensory actuator using an electrically conductive and photothermally responsive hydrogel, which combined the functions of piezoresistive strain/pressure sensing and photo/thermal actuation into a single material, as shown in Fig. 1.5g. Lee et al. [122] reported an approach to fabricate soft robots made of tensegrity structures using smart materials 3D printing combined with sacrificial molding, as shown in Fig. 1.5h. He et al. [123] built an electronically controlled soft gripper by using three tubular actuators based on the liquid crystal elastomer (LCE). Controlled by an externally applied electrical potential, the tubular actuator could exhibit multidirectional bending as well as large (about 40%) homogenous contraction, as shown in Fig. 1.5i. Xing et al. [124] designed a five-module super-lightweight soft manipulator that was driven by dielectric elastomer actuator (DEA). It had a length of 32 cm and a weight of only 18 g, as shown in Fig. 1.5j.

1.2.4 Shape Memory Materials-Driven Mode

The soft manipulator with shape memory materials driven mode gets the drive force from the shape restore of the materials at a certain temperature. When the temperature reaches a certain level, the shape memory material can restore its initial memorized shape. To make use of the shape restoration process, the drive force can be obtained. According to different materials, it is divided into shape memory alloy (SMA) and shape memory polymer (SMP). As reported in the literature, the SMA is used commonly in the shape memory materials-driven soft manipulator, as shown in Fig. 1.6.

Laschi et al. [21, 22, 138–140] designed and developed a soft manipulator imitating octopus tentacle based on the SMA-driven principle, as shown in Fig. 1.6a. It could perform the bending and curling movement through the electrical heating of the SMA springs, which were arranged along the transverse and longitudinal direc-

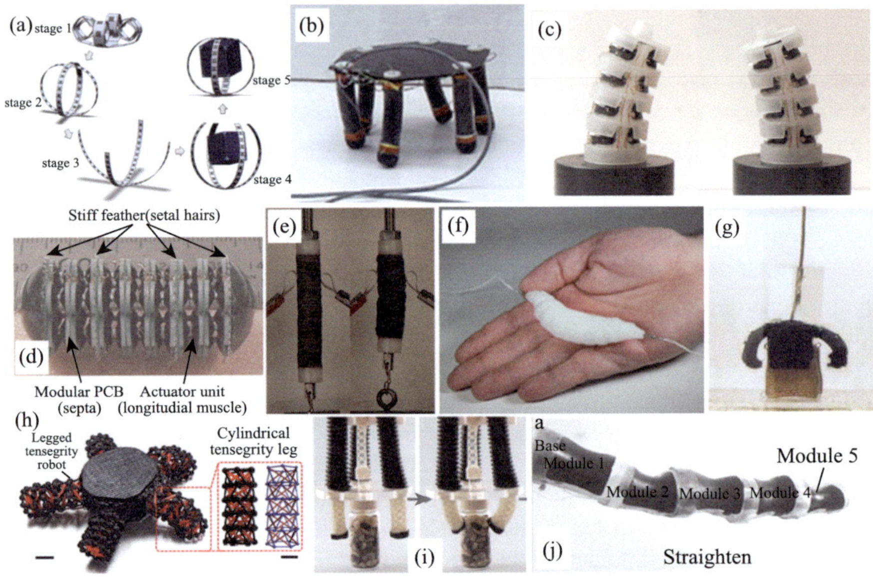

Fig. 1.5 Electroactive polymer-driven soft manipulator: **a** DEMES [117]: **b** spring roll [20]; **c** ANTLA [118]; **d** multi-stacked actuator [119]; **e** stack actuator [18]; **f** soft artificial muscle [120]; **g** somatosensory actuator [121]; **h** starfish robot [122]; **i** LCE-based tubular actuator [123]; **j** Soft manipulator driven by dielectric elastomers [124]

tions in the soft manipulator. Cheng et al. [134] designed and developed a minimally invasive neurosurgical intracranial robot (MINIR), which was composed of three actuators that were made up of SMA springs and the cooling mechanism, as shown in Fig. 1.6b. Zhang et al. [135, 141, 142] fabricated an octopus-like soft robotic arm by embedding the SMA springs into the silicone soft module, as shown in Fig. 1.6c. The motion of the soft arm was controlled by the coupling effect of the SMA springs and the silicone structure. Wang et al. [136] described a bioinspired soft robotic finger made up of the shape memory alloy-based soft actuator with embedded stiffness-varying structures serving as modifiable endoskeletons, as shown in Fig. 1.6d. Wang et al. [137] described a shape memory alloy-based soft gripper composed of three identical fingers with variable stiffness for adaptive grasping in low stiffness state and effective holding in high stiffness state, as shown in Fig. 1.6e.

1.2.5 Electromagnetic-Driven Mode

The soft manipulator with electromagnetic-driven mode conducts its motion control based on the interaction effect of the ferromagnetic soft material and the external magnetic field. The difficulty lies in the design of the magnetic soft robot and the

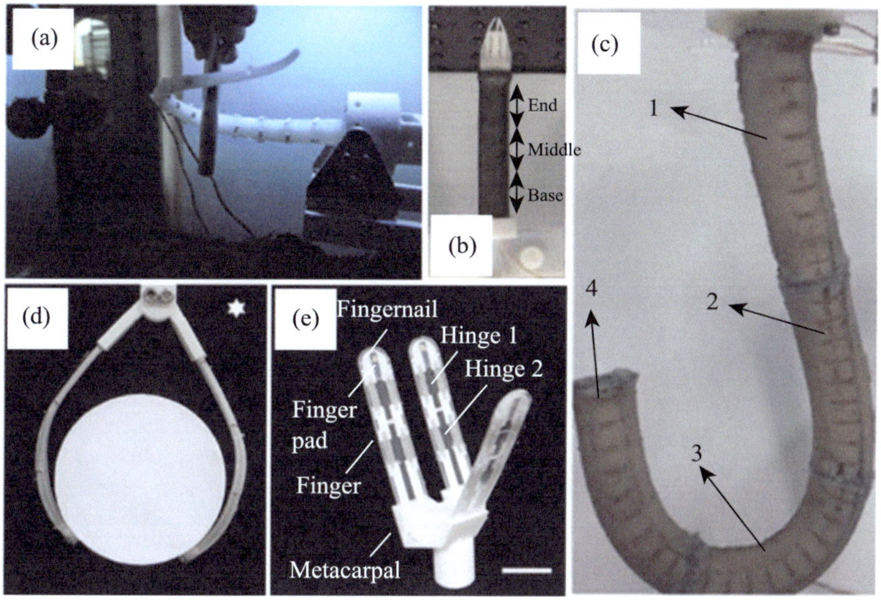

Fig. 1.6 Shape memory materials-driven soft manipulator: **a** bionic octopus tentacle [22]; **b** MINIR-II [134]; **c** soft robotic arm [135]; **d** shape memory alloy-based soft finger [136]; **e** shape memory alloy-based soft gripper [137]

development of the stable and controllable strong magnetic field environment. The electromagnetic-driven soft robot is made of the ferromagnetic soft materials by programmable magnetic printing or mixed casting. The external magnetic field is generated by using the electromagnetic coils distributed in a complex way. The typical design cases of the electromagnetic-driven soft robot are shown in Fig. 1.7.

Kim et al. [23] designed and developed a ferromagnetic soft continuum robot, which was used for minimally invasive blood vessel cleaning, as shown in Fig. 1.7a. It was invented based on their existing technologies such as programmable magnetic printing [24] and hydrogel bioelectronics devices [25]. Firstly, the micrometer-scale soft magnetic wire was printed and polarized by using soft magnetic materials. Then the soft wire was covered with a hydrogel skin of a micron thickness. Under the action of external magnetic field, the ferromagnetic soft robot could make spatial motions in a predetermined manner, and could quickly pass through a series of complex vascular-like channels with the diameter of only 2–3 mm. Choi et al. designed a magnetically controlled soft microrobot, which was composed of a soft microrobot, an electromagnetic generating and control system, and a camera monitoring system, as shown in Fig. 1.7b [26, 27, 147–149]. The soft microrobot was fabricated by molding a PMDS micro beam (1 mm diameter), a neodymium magnet, a connection (a microspring), and a guidewire together with the liquid silica gel. Under the action of the 15 mT external magnetic field, the maximum deflection of the microrobot can reach 132.7°. Kim et al. [143] introduced a strategy for encoding reprogrammable

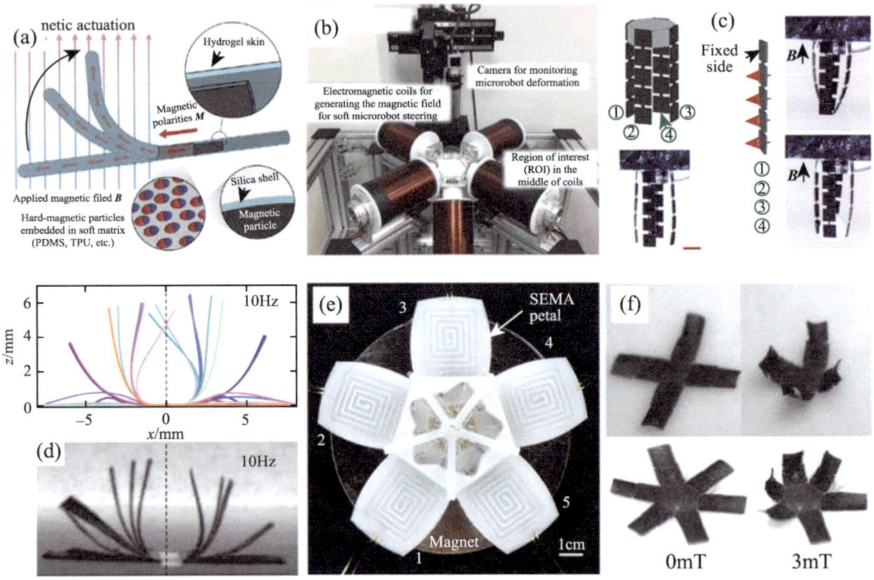

Fig. 1.7 Electromagnetic-driven soft robot: **a** ferromagnetic soft continuum robot [23]; **b** magnetically controlled soft microrobot [26]; **c** 3D-configured magnetic soft gripper [143]; **d** ultrafast soft robot [144]; **e** soft electromagnetic actuator (SEMA) [145]; **f** untethered multi-arm magnetic microgripper [146]

shape-morphing instructions into magnetic soft machines, as shown in Fig. 1.7c. Wang et al. [144] demonstrated a series of simulation-guided lightweight, durable, untethered, small-scale soft-bodied robots that were driven at low magnetic fields with large-degree deformations, as shown in Fig. 1.7d. Kaltenbrunner et al. introduced soft electromagnetic actuator (SEMA) by replacing solid metal coils with liquid-metal channels embedded in elastomeric shells and conducted the demonstration using a multicoil flower SEMA with individually controlled petals, as shown in Fig. 1.7e [145]. Xu et al. [146] fabricated untethered multi-arm magnetic microgripper by using a method for patterning hard magnetic microparticles in an elastomer matrix, as shown in Fig. 1.7f.

1.2.6 Hybrid Drive Mode

Hybrid drive mode is a combination of two or more different drive modes to improve the load capacity or the control accuracy of the soft manipulator. For example, one of the typical hybrid drive modes is the combination of the fluid-driven mode and the cable-driven mode. The fluid-driven mode has a good characteristic of pressure bearing, but a weak performance of tensile bearing. The cable-driven mode has strong tensile resistance, but cannot bear pressure. The combination of these two driven modes can give the soft manipulator good characteristics of pressure bearing and tensile bearing, thus increasing its load capacity and position control accuracy. The hybrid drive mode keeps the complementary advantages of different drive modes. However, it greatly increases the difficulties and technical challenges of the design and development of the control system as well as the control algorithm. The typical design cases of the hybrid drive mode are shown in Fig. 1.8.

Laschi et al. designed and developed a soft manipulator, which was composed of three McKibben pneumatic actuators that were connected by a series of plates, as shown in Fig. 1.8a [150, 158, 159]. The soft manipulator was controlled using the hybrid mode of the gas-driven mode and the cable-driven mode. Kang et al. [151, 160] presented a generic cable-driven structure, which could obtain a variable stiffness by a set of embedded SMA springs, as shown in Fig. 1.8b. Diteesawat et al. [152] presented an electro-pneumatic pump (EPP), which could silently control volume and pressure for soft robots, overcoming the limitations of existing pneumatic power sources, as shown in Fig. 1.8c. Takeuchi et al. [153] developed a biohybrid robot actuated by an antagonistic pair of skeletal muscle tissues, which achieved large actuation (about 90° of rotation of a joint) and a long lifetime (about 1 week), as shown in Fig. 1.8d. Acome et al. [154] described a class of hydraulically amplified self-healing electrostatic (HASEL) actuators, which harnessed a mechanism that coupled electrostatic and hydraulic forces to achieve a variety of actuation modes, as shown in Fig. 1.8e. Subramaniam et al. [155] presented a soft gripper with an active

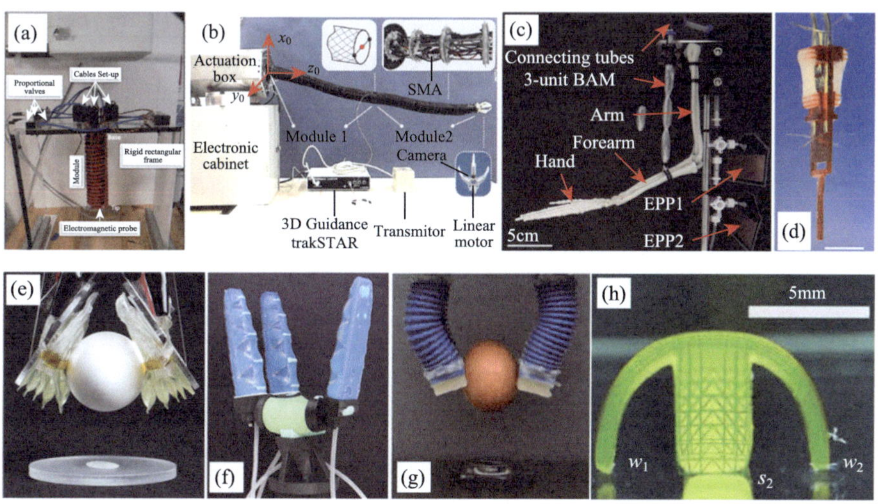

Fig. 1.8 Hybrid drive soft manipulator: **a** soft manipulator driven by fluidic pressure and cables [150]; **b** rod-driven structure embedded with SMA springs [151]; **c** electro-pneumatic pump (EPP) [152]; **d** biohybrid robot with an antagonistic pair of skeletal muscle tissues [153]; **e** hydraulically amplified self-healing electrostatic (HASEL) actuator [154]; **f** hybrid soft gripper [155]; **g** bistable hybrid soft gripper [156]; **h** programmable aquatic hydrogel robot [157]

palm to control grasp postures, which could grasp a wide range of part geometries and compliances while achieving a maximum payload of 30 N, as shown in Fig. 1.8f. Tang et al. [156] presented a bistable hybrid soft gripper with wide-range variable stiffness modulation, which was controlled through air pressure and spring pulling, as shown in Fig. 1.8g. Li et al. [157] reported a programmable aquatic hydrogel robot that changed its shape in response to light and experience torques in rotating magnetic fields, as shown in Fig. 1.8h.

1.2.7 Comparison of Different Drive Modes

From the typical cases of the soft manipulators with different drive modes, it is known that the soft manipulator has different performance when its drive mode is different. The fluid-driven mode, especially the gas-driven mode, has the advantages of light weight, quick response speed and large output torque. The liquid-driven mode is usually used to control the underwater soft manipulator. The disadvantage of the fluid-driven mode lies in that the pneumatic or hydraulic control system is complicated with a relatively big volume and a heavy weight. This gives a great difficulty to the miniaturization of the whole soft manipulator system and limits its application in some specific fields such as space engineering. The cable-driven mode can transmit the drive force from a long distance to ensure a small moment of inertia of the manipulator. However, it needs to be equipped with motors and cable retracting and releasing devices, which are also difficult to get miniaturized. The EAP-driven mode has the advantages of light weight, high energy density, and easy to get miniaturized. However, it is also facing the difficulties of the material damage caused by the high-level voltage, the weak load capacity, or the low control accuracy. The shape memory material driven mode can provide a relatively sufficient torque with a miniaturized structure. However, the phase changed by heating result in a slow response speed and a limited control pattern. These disadvantages limit its control accuracy and application fields. The electromagnetic-driven mode has a good prospect in the medical field. The body-part of the electromagnetic-driven soft robot can be miniaturized in millimeter scale, although it relies on a complex external equipment to generate a stable magnetic field for its motion control. The hybrid drive mode keeps the complementary advantages of different drive modes, which is helpful to improve the control accuracy and the load capacity of the soft manipulator. However, it takes great complexity and difficulty to the control system. The performance comparison of the soft manipulators with different drive modes is listed in Table 1.1. It can be concluded that the different drive modes have different advantages and disadvantages, the selection of the drive mode for the soft manipulator should jointly consider the specific task requirements and mission scenarios.

Table 1.1 The performance comparison of the soft manipulators with different drive modes

Drive mode		Response speed	Load capacity	Control accuracy	Additional control device	Miniaturization
Fluid-driven	Gas-driven	Quick	Large	Relatively high	Complicated	Difficult
	Liquid-driven	Quick	Large	Relatively high	Complicated	Difficult
Cable-driven		Quick	Large	Relatively high	Complicated	Difficult
Electroactive polymer driven	Electronic type	Quick	Medium	Relatively high	Simple	Easy
	Ionic type	Quick	Small	Low	Simple	Easy
Shape memory materials driven	SMA	Slow	Medium	Low	Simple	Easy
	SMP	Slow	Medium	Low	Simple	Easy
Electromagnetic-driven		Quick	Small	Relatively high	Complicated	Difficult
Fluid-cable hybrid drive		Quick	Large	High	Complicated	Difficult

1.3 Enabling Technologies of the Soft Manipulator

The design of the soft manipulator is inspired by the soft tissues such as elephant trunk and octopus's tentacles. Besides the design and fabrication, the enabling technologies of the soft manipulator are mainly distributed in the following aspects: kinematic and dynamic modeling, motion control, shape detection, dynamic tactile perception, stiffening method, as shown in Fig. 1.9.

As shown in Fig. 1.9, the kinematic and dynamic model is the theoretical way to investigate the characteristic of the soft manipulator, in which the piece-wise constant curvature (PCC) and the Cosserat geometrically exact model are widely studied in recent years. The kinematic and dynamic model is an important support to the model-based motion control of the soft manipulator. The model-based open-loop control is relatively traditional and widely studied, while the model-free and close-loop control are newly emerging. The shape detection and the dynamic tactile perception can provide the feedback of real-time shape and grasping force for the close-loop control of soft manipulator. The stiffening method is to change the stiffness of the soft structure actively or passively, to get a balance between the load capacity and the flexible locomotion. The enabling technologies as well as their properties and examples are listed in Table 1.2.

Table 1.2 Enabling technologies of the soft manipulator

Enabling technologies	Methods	Properties	Examples and references
Kinematic and dynamic modeling	Constant curvature method	Total constant curvature	BHA [66]; Soft robotic manipulator [161]; Foldable pneumatic soft manipulator [162]
		Piece-wise constant curvature	Multisection continuum robot [163]; Elephant's trunk manipulator [164]; Underwater soft manipulator [31]
	Non-constant curvature method	Super redundant continuum approximation method	Continuum space manipulator [108]
		Lumped-mass methodology	Soft continuum robot [165]
		Piecewise-constant strain	Soft manipulators with tendon and fluidic actuation [159]; BHA [30]
		Cosserat geometrically exact model	OctArm [56]; Multibending soft robot [100, 101]
Motion control of soft manipulator	Model-based control	Control based on the kinematics and dynamics model	PneuArm [9]; OctArm [58]; HPN [61, 62]; BHA [63]; Multi-section continuum manipulator [88]; Tendon-driven continuum robot [90]
	Model-free control	Control by means of vision, intelligent algorithm	HPN [60]; ASR [166]; Soft robotic endoscope [167]

(continued)

Table 1.2 (continued)

Enabling technologies	Methods	Properties	Examples and references
Shape detection technology	External detection method	Visual measurement	BHA [63]; Cable-driven soft manipulator [114]
	Embedded detection method	Cable measurement, optical fiber measurement	Soft actuator [71]; Soft robotic arm [107]; Cable-driven soft manipulator [111] Soft pneumatic actuator [168]; Proprioception with optical lace [169]; Fiber optic shape sensor [170]
	Integrated strain sensors	Printing self-coagulating conductive pickering (SCP) emulsion onto soft robots	Soft gripper integrated SCP emulsion [171]; Soft robot perception using embedded soft sensors [71]
Dynamic tactile perception	Flexible tactile sensor	Resistance, capacitance, piezoelectric	Soft pneumatic gripper [172]; Soft pneumatic gripper (SPG) [173]; Soft magnetic skin [174]; Biosensing soft gripper [175]; Flexible textile tactile sensors [176]
Stiffening method	Jamming-based stiffening	In-series connection of the granular jamming membranes	Multi-module manipulator [67]
	Low melting point alloy (LMPA)	Phase transformation by melting the metal via joule heating	Soft robotic gripper [177, 178]; Variable stiffness actuator [179]
	SMA and SMP	Soft composite actuator with stiffness changeable material	Leverage mechanism for stiffness adjustment [151]; Stiffness-varying structure embedded with SMP [136]; Shape memory actuators of stiffness modulation [137]

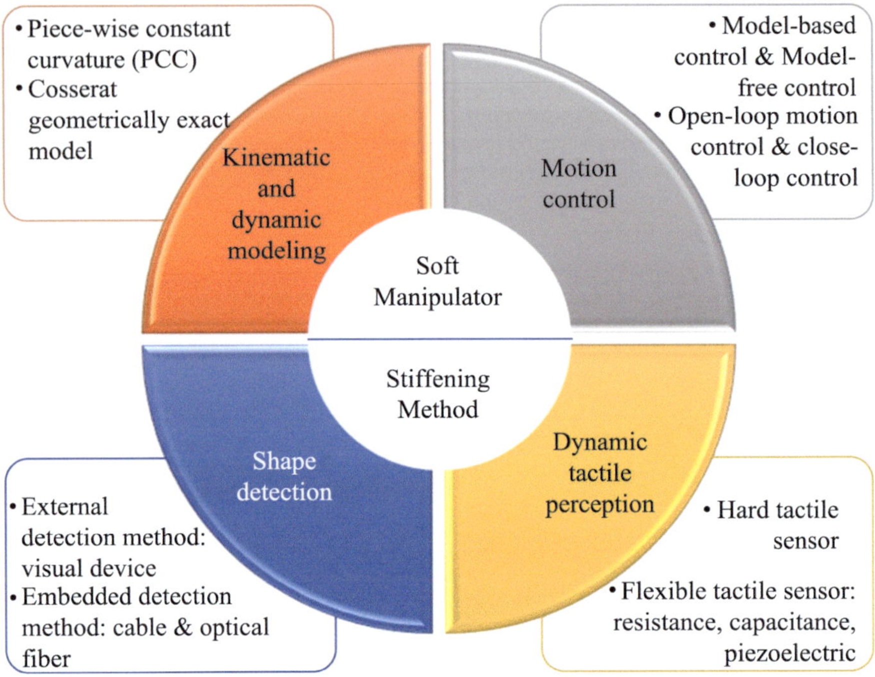

Fig. 1.9 Architecture of enabling technologies of the soft manipulator

1.3.1 Kinematic and Dynamic Modeling

The kinematic and dynamic modeling of the soft manipulator needs to consider the highly nonlinear mechanical properties because of its soft materials, flexible structure and special actuation [180]. The kinematics modeling methods are mainly divided into the constant curvature method and the non-constant curvature method. The constant curvature method is to solve the modeling problem by approximating the soft manipulator to an arc with constant curvature. This method is simple and easy to implement and has been widely used in the early stages of theoretical research on the soft manipulator [181, 182]. However, with the increasing complexity of the design configuration and external environment load of the soft manipulator, the assumption of global constant curvature is too simple to be applied. In order to reduce the model error, the piece-wise constant curvature (PCC) is proposed based on the constant curvature assumption [164, 182]. The solution is to divide the soft manipulator into several segments and establish the kinematic model in each segment by using the constant curvature assumption. The overall kinematics model of the soft manipulator can be established by assembling the piece-wise kinematic models of all segments [181, 183]. The PCC method is simple and efficient, and can be applied to more complex problems, which greatly promotes the kinematics modeling theory of the soft manipulator [161, 162]. However, it is still unable to avoid the model error caused by the constant curvature assumption. This method has an inherent defect in its theoretical foundation, which restricts the improvement of model prediction accuracy. In order to make up for the theoretical defects of the constant curvature method, the non-constant curvature method comes into being.

The non-constant curvature method, as the name implies, is to abandon the assumption of the constant curvature and no longer approximately equates the whole or local part of the soft manipulator into a constant curvature arc. It is mainly implemented by using the following methods: the super redundant continuum approximation method, the spring mass model, the Cosserat geometrically exact model, the quasi-static analytical model and so on [17]. The super redundant continuum approximation method was firstly proposed in the 1990s, and its solution is to use a continuous curve to approximately describe the geometric deformation characteristics of the manipulator [184]. It is suitable for the super redundant manipulator with a large number of control freedoms, such as the snake-like manipulator or the trunk-like manipulator [184–186]. The spring mass model is to divide the soft manipulator into several segments, in which the spring mass equivalent model is used to describe its kinematic characteristics. The overall kinematic characteristics of the soft manipulator are obtained by assembling the models of all the segments. This method is only applicable to certain types of soft manipulators, such as the pneumatic trunk-like soft manipulators or the shape memory spring driven soft manipulators [161]. In recent years, the Cosserat geometrically exact model is applied to solve the kinematic modeling of the soft manipulator [187]. In this solution, the soft manipulator is firstly divided into several segments. In each segment, the kinematic model is established based on the Cosserat elastic beam assumption, which is more accurate by considering both the tensile deformation and the lateral deflection, as is shown in Fig. 1.10. This method has been applied to the kinematics modeling process of octopus tentacle like soft manipulator [17, 188], OctArm series soft manipulator [57] and cable-driven backbone continuum manipulator [189, 190]. Most of these soft manipulators adopt variable cross-section configurations, which makes the constant curvature method unable to be applied. Therefore, the non-constant curvature method must be explored. Currently, the Cosserat geometrically exact model gives a good solution to break through the constraint of constant curvature assumption, and agrees well with the real physical deformation of the soft manipulator. It can improve the prediction accuracy of the kinematic model, thus has a wide range of application. In the solutions of the quasi-static analytical model, Olson et al. [191] presented a quasi-static analytical model of soft arms bent with longitudinal actuators, which is presented as a generalizable framework and specifically implemented for an arm with N fluid-driven actuators. Tutcu et al. [192] developed a quasi-static model coupled with a kinematic model to accurately position the end effector while accounting for the inflated beam stiffness and end-effector loads. Shiva et al. [193] proposed a quasistatic approach for 3D modeling and real-time simulation of a pneumatically actuated soft continuum robotic appendage to estimate the contact force and overall pose. Additionally, the researchers have proposed other novel kinematic modeling methods according to their own situations. Shan et al. [194] presented a detailed mathematical modeling and performance analysis of parallel grippers equipped with soft fingers taking advantage of the fin ray effect (FRE). Sedal et al. [195] studied models for the fiber-reinforced soft actuator and those are used in several soft articulated systems, including soft manipulators. Jeong et al. [196] presented a framework for the reliability analysis of soft robotic devices and examined a case study of a tendon-driven soft robot. Hamaya et al. [197] proposed an approach for obtaining accurate soft actuator models through the design of physical user-robot interactions for wearable robots. Morales et al. [93] presented a numerical model of a deformable manipulator based on the finite element method (FEM), using a set of beam elements that reproduce the lattice structure of the robot. Moreover, other quantitative modeling methods are also tried to solve the kinematic modeling of the soft manipulator, for example, Singh et al. [198] proposed a quantitative modeling method based on the Pythagorean hodograph (PH) curves. The aim was to obtain a three-dimensional reconstruction of the shape of the continuum manipulator with variable curvature, allowing the calculation of its inverse kinematic model (IKM). Sadati et al. [199] introduced two modeling methods for continuum manipulators, including a general reduced-order model (ROM) and a discretized model with absolute states and Euler-Bernoulli beam segment (EBA). Camp et al. [200] proposed a fixed-end cylindrical model to predict the strain of a McKibben fluidic artificial muscle (FAM). Hussain et al. [116] presented a mathematical formulation based on the screw theory to model a tendon-driven soft gripper. Digumarti et al. [201] presented a method based on elliptic Fourier descriptors to describe soft deformable morphologies and performed eigenshape analysis on the descriptors to extract key features that changed during the motion of soft robots. Yang et al. [151] proposed a geometric constraint-based model to describe the deformation and mechanics of a rod-driven continuum robot.

Based on kinematics description, the dynamic modeling of soft manipulator needs to further consider the external load, such as the dynamic contact force in the process of target grasping and the external environment disturbing force. The external loads take great difficulties in the modeling and control of the soft manipulator. Currently, the researches on the soft manipulator are in the primary stage in both theoretical and engineering aspects. The theoretical research is mainly focused on the kinematic modeling, while the dynamic modeling problems are catching more and more attention. For example, Luo et al. [202] addressed a general framework to solve the 2D modeling problem of a pressure-operated soft snake robot, which was applicable to most pressure-operated soft robots developed by a modular kinematic arrangement of bending-type fluidic elastomer actuators. Mustaza et al. [203] proposed a visco-hyperelastic material-based continuum mechanics approach for the dynamic modeling of fiber-reinforced softmanipulators.

1.3.2 Motion Control

The motion control of the soft manipulator is mainly realized by the deformation accumulation of its soft structure, which is completely different from that of the rigid

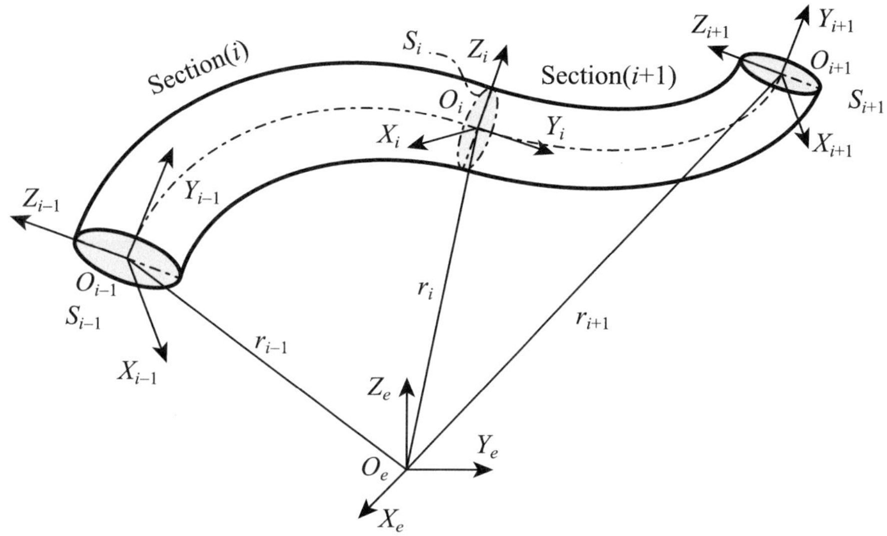

Fig. 1.10 Principle of the Cosserat geometrically exact model

manipulator. The motion control is to establish the transfer relationship between the control variables and the shape deformation of the soft manipulator. According to whether it depends on the theoretical model, the control can be classified into the model-based control and the model-free control [204].

Model-based control is to obtain the transfer function relationship between the input variables and the output parameters of the control system based on the kinematics and dynamics model, and then carry out the control algorithm design [205]. The input variables of the soft manipulator control system are mainly determined by the drive mode. For example, the input variables in fluid-driven mode are air pressure or hydraulic pressure, the input variables in cable-driven mode are cable length, and the input variables in electromagnetic-driven mode and intelligent material-driven mode are voltage or current. The difference of input variables will lead to great difference of transfer function relationship, which leads to the difference of the model control algorithm design. Currently, the model-based control has been widely used in fluid-driven mode and cable-driven mode. In the control of fluid-driven mode, a typical application was the control of the trunk-like soft manipulator [30, 72, 75]. The kinematics model was firstly established by the PCC method to obtain the Jacobian matrix, and then the open-loop motion control of the soft manipulator was carried out. Otherwise, Trumić et al. [206] introduced closed-loop control approach comprising an adaptive compensator and a dynamic decoupler by introducing a first-order dynamic equation for the stiffness variable and using the additional control degree of freedom, embedded in the null space of the pneumatic actuator matrix. Gong et al. [31] performed a real-time closed-loop control of the spatial location and trajectory of an underwater soft manipulator based on a simple and efficient inverse kinematics model. Luo et al. [207] presented a sliding-mode controller formulation considering a modular soft robotic architecture that offered proprioceptive sensing of pressure-operated bending actuation modules. In the control of cable-driven mode, a typical application was the control of the soft manipulator imitating octopus' tentacle [99, 101, 112]. On the basis of the kinematic model, the functional relationship between cable length and shape deformation was established, and the control algorithm design and experimental verification were carried out. Moreover, other model-based control methods were studied by the researchers. Amanov et al. [90] validated a model-based control for a tendon-driven continuum robot, which was designed on a small scale with diameters of below 10mm. Pozzi et al. [208] investigated the grasps with soft hands according to a model-based approach, in which the so-called closure signature was used to model closure motions of soft hands by associating to them a preferred grasping direction. Morales et al. [93] investigated a closed-loop control strategy for a deformable manipulator, which was implemented to account for the disparities between the model and the robot. Greer et al. [209] developed a method to plan paths for a soft growing robot based on an interaction model with a planar environment in an empirical kinematic model. Tang et al. [210] proposed a model based online learning and adaptive control algorithm for the wearable soft robotic glove, taking its interaction environment into account. Franco et al. [211] investigated the control problem for soft continuum manipulators that are subjected to unknown disturbances acting in the plane of bending based on an energy-shaping approach. Best et al. [43] developed a sliding mode controller and model predictive controller based on a model that includes stiffness as a state variable, which could control stiffness and manipulators simultaneously. Zhang et al. [212] proposed a kinematic obstacle avoidance algorithm for space hyper-redundant manipulators using a static and dynamic curve to constrain the shape of the manipulatorssimultaneously.

Compared with model-based control, model-free control refers to the motion control of the soft manipulator without depending on the specific physical model or only relying on simple physical model. It has wide and general applicability because it has no special requirements for the control objects. However, it needs to carry out abundant learning and training based on the prototype system. Compared with the certainty and stability of the rigid manipulator, there are more uncertainties in the motion control of the soft manipulator, and the model-free control is more challenging in the technical implementation. Despite these existing difficulties, the model-free control still attracts a great research interest. For example, You et al. [60] studied the planar motion control of the soft manipulator driven by the honeycomb pneumatic network through the reinforcement learning method. Yip et al. [213, 214] established the Jacobian matrix prediction model through the online environment interaction to realize the motion control of a cable-driven continuum manipulator. Cherubini et al. [215] proposed heuristic methods and a neural network for mapping the visual state to human actions. Mengacci et al. [166] controlled an articulated soft robot by using the model-free torque iterative learning control. Lee et al. [167] proposed a generic control framework based on the nonparametric, online and local training to learn the inverse model directly, without prior knowledge of the robot's structural parameters. Zhang et al. [216] proposed an adaptive stiffness control strategy in order to improve the robot's physical stability in complex environments. Khan et al. [217] presented the comparative performance of several variants of model-free PID-controllers based on extensive experimental results. They presented a heuristic-based coordinate descent algorithm to tune the controller parameter automatically. Li et al. [218] developed a model-free method based on an adaptive Kalman filter to accomplish path tracking for a continuum robot using only pressures and the tip position.

The control of the soft manipulator has a close relationship with the design and fabrication of the soft manipulator, which is still in the primary exploration stage. Currently, the main control method is open-loop motion control with little consideration of the external load or environmental interaction. The research on dynamics modeling as well as its closed-loop control need further development. Meanwhile, the model-based control still faces technical challenges. For example, the prediction accuracy of the constant curvature model needs to be improved, the singularity often appears in the inverse solution of the non-constant curvature model, and the real-time shape detection and feedback need to be improved [183, 219]. The model-free control is still faced with great constraints of the low technology maturity of the soft manipulator system.

1.3.3 Shape Detection

The shape detection technology can provide real-time shape feedback for the autonomous motion planning and the closed-loop control of the soft manipulator. Considering the installation inside or outside the soft structure, the shape detection method can be classified into two types: the external detection method and the embedded detection method. The external detection method mainly uses visual measurement, which has been widely used in the prototype test and model verification of the soft manipulator. For example, Mcmahan et al. [10] used the 3D vision measurement system to obtain the real-time deformation and relative position information for the motion control of OctArm. Wang et al. [13] used the VICON visual motion capture system to get the real-time shape deformation of the cable-driven continuum manipulator, then to verify the prediction accuracy of the proposed model. Huang et al. [108] obtained the shape change of the cable-driven soft manipulator by the visual method, which was used to verify the prediction accuracy of the kinematic model. Wang et al. [112] also used the VICON visual motion capture system to obtain the spatial motion of the tentacle-like soft arm, which was used to verify the accuracy of the proposed three-dimensional kinematic model. Liu et al. [162] obtained the shape deformation of the foldable soft manipulator using the visual measurement method and verified the prediction accuracy of the proposed kinematic model.

The external detection method based on the visual measurement is relatively mature and reliable but is difficult to be applied in the narrow or unstructured environment. In such an application scenario, the soft manipulator must be equipped with an embedded detection device to get its real-time shape perception. The solution of the embedded detection method is to embed sensors and other measuring devices into the soft manipulator. The shape deformation of the soft manipulator will cause the real-time signal change of the sensors. By establishing the transferring model between the signal change and the shape deformation, the real-time shape of the soft manipulator can be obtained. In recent years, the embedded detection methods have been studied widely by using the cable measurement and the optical fiber measurement. The cable measurement is to calculate the shape deformation of the soft manipulator by arranging several cables along its central axis and measuring the cable length variations. The PCC modeling method is always used in this calculation. The cable measurement has been successfully applied in the OctArm soft manipulator [11] and the elephant trunk-like soft manipulator BHA [63, 64]. The optical fiber measurement is to calculate the shape deformation by embedding a set of distributed fiber Bragg grating (FBG) sensors in the soft manipulator and transforming the node strain of the FBG sensor network into the shape change based on the assumption of PCC, as shown in Fig. 1.11. This method has been successfully applied in the real-time shape detection of the pneumatic artificial muscle [107], the cable-driven micro soft manipulator [220, 221], and the octopus tentacle-like soft manipulator [111]. All these above indicate that the cable measurement is a relatively traditional solution compared with the optical fiber measurement. The cable measurement needs to be equipped with a complex cable retraction and length counter devices. Moreover, the cables are arranged along the outer surface of the soft manipulator with maintaining a certain tension, which produces a great additional impact on the motion control of the soft manipulator. The FBG sensors are directly embedded in the soft manipulator and produce little disturbance to the motion control of the softmanipulator.

Currently, the embedded shape detection method of the soft manipulator is still in the primary stage of exploration. Although the optical fiber measurement shows good application prospects, it is still facing technical challenges such as being easily disturbed by the external environment and the measurement system needing to be optimized. To avoid these technical challenges, some novel shape detection methods have been investigated by the researchers, as shown and listed in Fig. 1.12.

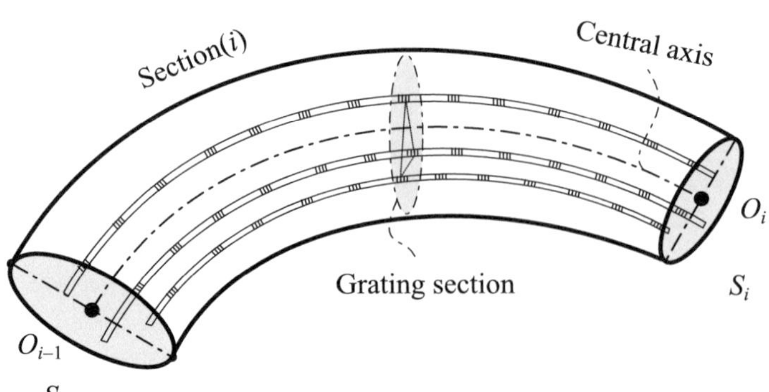

Fig. 1.11 Principle of the optical fiber measurement

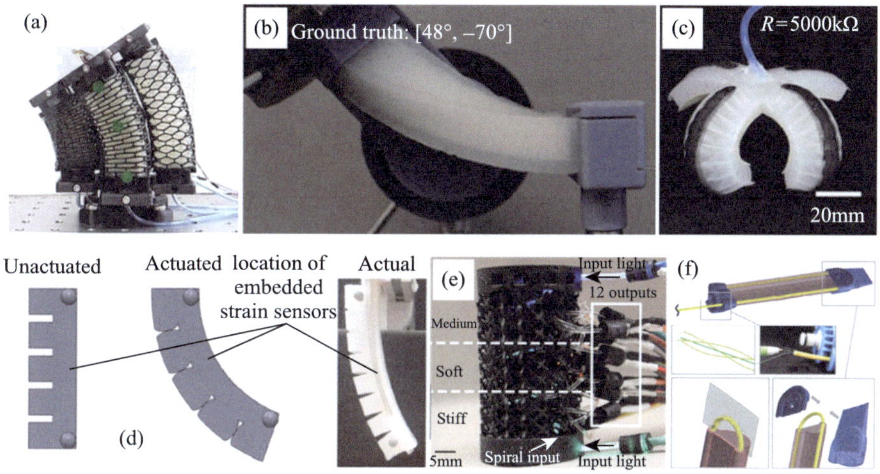

Fig. 1.12 Shape detection methods: **a** multifunctional soft sensor for pneumatic actuators [222]; **b** poprioception with soft optoelectronic sensory foams [223]; **c** soft gripper integrated strain sensors [171]; **d** soft robot perception using embedded soft sensors [71]; **e** proprioception with optical lace [169]; **f** fiber optic shape sensor (FOSS) [170]

As shown in Fig. 1.12a, Kim et al. [222] reported a multifunctional soft sensor capable of decoupling combined deformation modes of stretching, bending, and compression, as well as detecting individual deformation modes, in a compact form factor. van Meerbeek et al. [223] presented an internally illuminated elastomer foam that had been trained to detect its own deformation through machine learning techniques, as shown in Fig. 1.12b. Kim et al. [171] monitored the movement of a soft gripper by printing the self-coagulating conductive Pickering (SCP) emulsion onto its surface, as shown in Fig. 1.12c. Thuruthel et al. [71] proposed a synthetic system using a redundant and unstructured sensor topology embedded in a soft actuator, which was inspired by the human perceptive system, as shown in Fig. 1.12d. Shepherd et al. proposed a proprioception method based on a sensing network using lacings of stretchable optical fibers distributed throughout 3D-printed elastomer frameworks, as shown in Fig. 1.12e [169]. Galloway et al. [170] integrated fiber optic shape sensor (FOSS) into the structure of a fiber-reinforced soft actuator to detect its shape, as shown in Fig. 1.12f.

1.3.4 Dynamic Tactile Perception

The dynamic tactile perception is to sense the external environment stimuli just like the soft biological tissues such as the elephant trunk or the octopus' tentacles. It is still a great challenge, but is essential for a soft manipulator to interact with its surroundings [224, 225]. In the primary stage, the hard tactile sensors, such as the force/torque sensor and the force sensitive resistance sensor, are tried to conduct the tactile perception. With the fast development of the flexible electronics, the flexible tactile sensors, also known as the flexible electronic skin or the bionic skin, play a dominant role in the tactile perception of the soft manipulator [226]. The flexible sensors, which are attached to the external surface, can help the soft manipulator interact with the external environment by converting the environmental pressure stimulation into electrical signals. Compared with the hard tactile sensor, the flexible tactile sensor is much lighter and softer and can be made into various shapes and attached to the irregular surface of the soft manipulator. The flexible tactile sensor is generally composed of the electrode, dielectric material, active material, and flexible substrate. The active material layer, which is responsible for quantitatively converting the external environmental stimulus into detectable electrical signals, is the most important part of the tactile sensor. The dielectric material is usually located on both sides of the active material layer to receive and transmit electrical signals. The flexible substrate plays the role of carrying the tactile sensor and biocompatibility with biological tissues. The manufacture and material selection of the tactile sensor faces a great challenge to bear frequent tensile and compressive deformation. According to the type of the active material layer, the flexible tactile sensors can be classified as the resistance type, the capacitance type, and the piezoelectric type [227–229]. The resistance type is to perceive the environmental load by measuring the resistance value changes using the metal or semiconductor material. The capacitance type is to sense the pressure by measuring the capacitance changes between two parallel plates. The piezoelectric type is to conduct the tactile perception by using the piezoelectric material to generate charge (electric dipole moment) under the external stimulus.

Currently, the resistance type is the most popular choice for the tactile perception because it is more mature in technology and there are plentiful commercial products. For example, Din et al. [230] designed and developed a set of multi-mode tactile sensor systems, which could simultaneously measure the unidirectional normal stress, shear stress and pressure by embedding the copper resistance wire into the silicone substrate. Guo et al. [172] embedded the resistance wire into the soft silica gel on the outer surface of the soft gripper by photolithography, so as to realize the measurement of tangential stress during the grasping process. Preechayasomboon et al. [231] used the tactile sensor and conductive liquid to realize the measurement of the external pressure on soft structure and perception of its own length change. Suh et al. [232] measured static stress and dynamic stress at multiple points simultaneously by adopting both resistance type and piezoelectric type flexible sensors. Yildiz et al. [233] used microfluidic and flexible sensors embedded in the soft structure to measure the strain of the soft structure by measuring the change of resistance. Chen et al. [173] used a manometer to measure the air pressure charged into the soft gripper to realize the indirect measurement of the grasping force. Although the resistance type has been used widely in the tactile perception, it is still difficult to adapt to the large deformation of the soft manipulator. Facing these technical challenges, the improvements of the dynamic tactile perception have been studied by the researchers, as shown in Fig. 1.13.

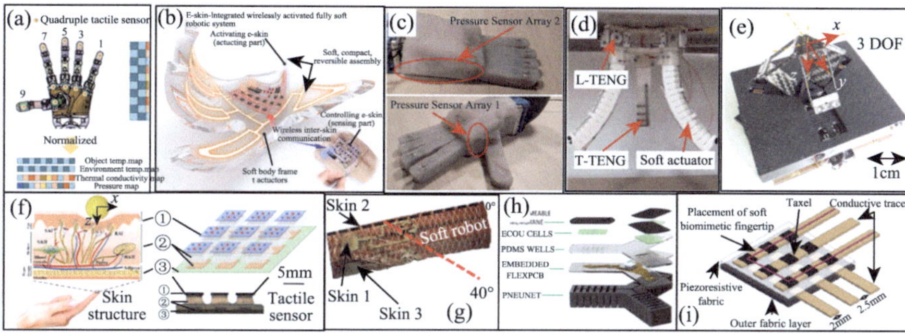

Fig. 1.13 Dynamic tactile perception: **a** Quadruple tactile sensor [234]; **b** E-skin-integrated wirelessly activated fully soft robotic system [235]; **c** silicon glove with embedded sensors [236]; **d** soft actuator integrated with TENG sensors [237]; **e** foldaway haptic interfaces [238]; **f** soft magnetic skin [174]; **g** stretchable elastic synaptic transistors [239]; **h** biosensing soft gripper [175]; **i** flexible textile tactile sensors [176]

As shown in Fig. 1.13a, Li et al. [234] reported the integration of quadruple tactile sensors onto a robot hand to precisely recognize different shapes, sizes, and materials in a diverse set of objects. Byun et al. [235] presented a skin-like electronic system that enabled a class of wirelessly activated fully soft robots whose driving part could be softly, compactly, and reversibly assembled, as shown in Fig. 1.13b. Mura et al. [236] presented a sensitized silicone glove in conjunction with a lightweight and impedance-controlled robot arm to investigate the role of arm stiffness as well as the kinesthetic synchronization of human and robot arm motions during the handshake, as shown in Fig. 1.13c. Jin et al. [237] reported a smart soft-robotic gripper system based on triboelectric nanogenerator sensors to capture the continuous motion and tactile information for the soft gripper, as shown in Fig. 1.13d. Mintchev et al. [238] proposed and demonstrated a portable three-degrees-of-freedom (3DoF) force feedback origami robot for human-robot interactions, as shown in Fig. 1.13e. Yan et al. [174] presented a soft tactile sensor with self-decoupling and super-resolution abilities by designing a sinusoidally magnetized flexible film whose deformation could be detected by a hall sensor according to the change of magnetic flux densities under external forces, as shown in Fig. 1.13f. Yu et al. reported a stretchable synaptic transistor based on elastomeric electronic materials, which was demonstrated in a soft adaptive neurologically integrated robot, as shown in Fig. 1.13g [239]. Justus et al. [175] reported a biosensing soft gripper featured FlexPCB and embedded phototransistors, which enabled the gripper to detect chemical signals by applying pressure and releasing the contents of a chemical-infused hydrogel, as shown in Fig. 1.13h. Sankar et al. [176] presented a pneumatically actuated soft biomimetic finger integrated with a textile neuromorphic tactile sensor array for a texture discrimination task. The tactile sensor outputs were converted into neuromorphic spike trains, which emulated the firing pattern of biological mechanoreceptors, as shown in Fig. 1.13i. All above introductions indicate that the flexible tactile sensor is widely used in biological, medical, intelligent wearable devices and other fields, and it will be increasingly highlighted in the dynamic tactile perception of the soft manipulator.

1.3.5 Stiffening Method of Soft Structure

The stiffening method is to change the equivalent stiffness of the soft structure actively or passively in a certain range. When it needs to capture an object, the soft manipulator gets a higher stiffness to improve its load capacity. When it needs to get into the narrow or unstructured environment, the soft manipulator gets a lower stiffness to become flexible. This is just the value and significance of the stiffening method. According to whether there is an external excitation, the stiffening method can be classified into the passive stiffening method and the active stiffening method.

The passive stiffening method is to change the stiffness by an origami structure design or internal friction of the granular jamming or the multi-layer structure. For example, Falco et al. [67] filled particles along the central axis of the soft structure and achieved variable stiffness through mutual extrusion between the air cavity and the particles. Kim et al. [240] got a variable stiffness of minimally invasive surgical soft catheter through the extrusion friction between multilayer composite structures. Mintchev et al. [241] demonstrated a dual-stiffness origami gripper that could be used for rigid grasping but softened to avoid overloading of the manipulated objects, as shown in Fig. 1.14a. Brancadoro et al. [242] proposed a fiber jamming stiffening mechanism and tried to analyze the main parameters influencing the fiber jamming transition as promising stiffening solution for soft robotics, as shown in Fig. 1.14b. Al Abeach et al. [243] designed a soft gripper with a variable stiffness by using two types of McKibben muscles, which were arranged to act antagonistically, as shown in Fig. 1.14c.

The active stiffening method is to change the stiffness by the external excitations, such as the fluidic-pressure/cable-tension excitation, the electric/magnetic field excitation, the temperature excitation and so on. The fluidic-pressure/cable-tension excitation refers to changing the stiffness by controlling the fluidic-pressure and the cable-tension independently or simultaneously. For example, Mahl et al. [30] increased the stiffness of the trunk-like soft manipulator by increasing the internal pressure. Laschi et al. [244, 245] designed a composite structure of the transverse and the radial pneumatic artificial muscles, and got a variable stiffness by limiting the local deformation of the radial muscle through the transverse muscle. Tang et al. [156] presented a design of a soft gripper with a wide range of tunable stiffness by using hybrid actuation of tendon-spring and air pressure, as shown in Fig. 1.14d. Similarly, Shahid et al. [246] developed a soft composite finger with tunable joint stiffness by using hybrid actuation of the cable tension and the air pressure, as shown in Fig. 1.14e.

The electric/magnetic field excitation refers to changing the stiffness by changing the viscidity and the modulus of the filled electrorheological or magnetorheological fluid using an external electric or magnetic field [247]. For example, Majidi et al.

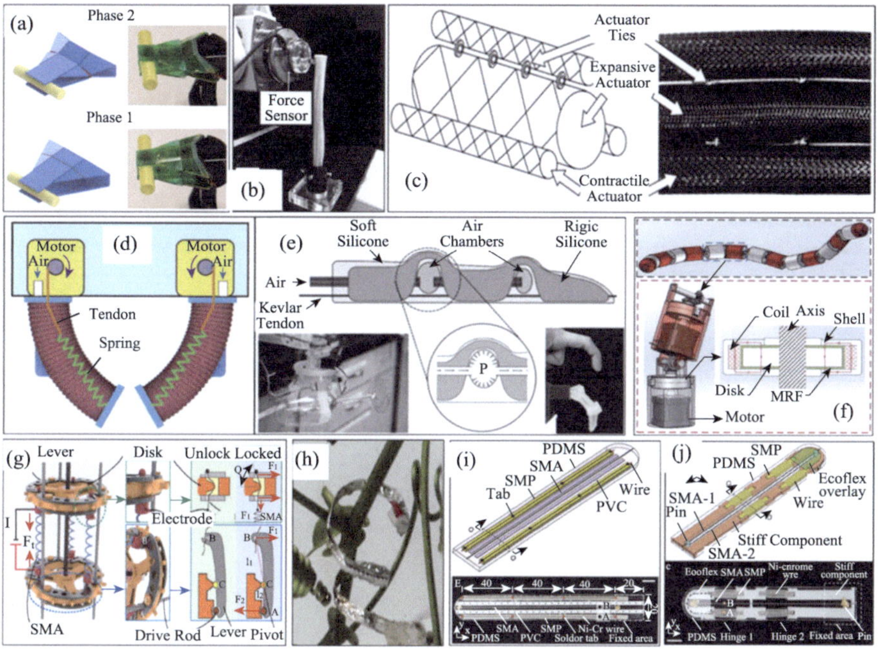

Fig. 1.14 The stiffening methods: **a** dual-stiffness origami gripper [241]; **b** fiber jamming transition as a stiffening mechanism [242]; **c** variable stiffness continuum finger [243]; **d** Bistable hybrid soft grippers [156]; **e** soft composite finger with tunable joint stiffness [246]; **f** the snake-like robot with variable stiffness actuators [216]; **g** leverage mechanism for stiffness adjustment (SMA) [151]; **h** a variable-stiffness tendril-like soft robot [254]; **i** stiffness-varying structure embedded with SMP [136]; **j** shape memory actuators of stiffness modulation [137]

[248] got a variable stiffness by putting a micro channel filled with magnetorheological fluid in a low electromagnetic field. Pettersson et al. [249] improved the load capacity of a gripper by filling the soft structure with the magnetorheological fluid to get a variable stiffness. Sadeghi et al. [250] gave a variable stiffness to a cable-driven soft actuator by filling the soft structure with electrorheological fluid and controlling the forward and reverse flow. Zhang et al. [216] proposed a variable stiffness actuator filled with magnetorheological fluid (MRF), as shown in Fig. 1.14f. The electric/magnetic field excitation can get a variable stiffness in a fast response. However, the equipment is so complex that there are a series of safety problems.

The temperature excitation refers to changing the stiffness by making the phase changing of the low melting point alloy or the shape memory materials happening in a special temperature range. For example, Hao et al. [177, 178, 251] embedded a low melting point alloy layer at the end of the soft actuator and changed its stiffness by heating the alloy to make phase transformation. When the alloy was heated to the liquid state, the soft actuator obtained flexible motion characteristics. When the alloy was cooled to the solid state, the stiffness of the soft actuator was strengthened.

Chautems et al. [252] used low melting point alloy materials to obtain a variable stiffness for a magnetic controlled cardiac catheter. Alambeigi et al. [253] made a cable-driven soft manipulator to obtain variable stiffness characteristics by using low melting point alloy materials. Shintake et al. [179] achieved the variable stiffness target grasping of a soft gripper by using both the low melting point alloy and the dielectric elastomer. Yang et al. [151] presented a novel variable stiffness mechanism powered by a set of embedded shape memory alloy (SMA) springs, which could make the drive rods become "locked" on the body structure with different configurations, as shown in Fig. 1.14g. Must et al. [254] demonstrated a variable-stiffness tendril-like soft robot based on the reversible osmotic actuation, as shown in Fig. 1.14h. Wang et al. [136] designed a stiffness-varying structure that was composed of shape memory polymer (SMP) embedded with nichrome (Ni-Cr) wires as heating elements, as shown in Fig. 1.14i. Wang et al. [137] described a soft finger with variable stiffness for adaptive grasping, which was fabricated by integrating a soft composite actuator with stiffness changeable material (SMA and SMP), as shown in Fig. 1.14j. The temperature excitation can obtain a wide range of variable stiffness. However, the heating and cooling time is generally long, which leads to a slow response of the stiffening process.

1.4 Opportunities and Challenges of Soft Manipulator in Application

The above research status and enabling technologies analysis indicate that the soft manipulator has a broad application modes and prospects. For different application fields, the selection of the design scheme and drive mode of the soft manipulator needs to be analyzed according to the specific requirements. The fluid-driven and cable-driven soft manipulators have the advantages of large acting torque and fast response speed, but they are faced with the difficulties of system miniaturization. The electromagnetic-driven soft manipulator is generally used in the medical field, which needs the complex external electromagnetic field environment. The soft manipulator driven by smart materials is easy to be miniaturized because it is driven directly by voltage or current. However, it also faces technical difficulties such as the enhancement of action torque and the improvement of control accuracy. Currently, the application of the soft manipulator lies mainly in the fields of the ground rescue, the underwater grabbing, the medical operation, the assist maintenance, the man-machine interaction, the space manipulation and so on, as listed in Table 1.3.

The ground rescue is to conduct inspection operations in the hazardous area using robotics technologies, which can increase the efficacy while decrease the risks [255–257]. As listed in Table 1.3, it can be known that in the application field of the ground rescue, the cable-driven mode and the fluid-driven mode are the main choices for the soft manipulator. In the application field of the underwater grabbing, the fluid-driven mode is undoubtedly the best choice for the soft manipu-

Table 1.3 Application fields of soft manipulator

Application field	Application scenario	Type of soft manipulator	Driven mode
Ground rescue	Inspections operations in hazardous area using unmanned remote handling systems	Long-range inflatable arm [102]	Cable-driven mode
		OctArm [10]	Fluid-driven mode
		Pressure-driven lengthening [34]	Fluid-driven mode
Underwater grabbing	Collecting in shallow water for replacing human divers	OBSS soft manipulator [31]	Fluid-driven mode
		Nanofiber-reinforced soft robotic actuators [47]	Fluid-driven mode
		Soft robots consisting of dual-morphing M-ori [36]	Fluid-driven mode
		Underwater robotic manipulator [46]	Fluid-driven mode
		Bionic octopus tentacle [22]	Shape memory materials-driven mode
Medical operation	Take advantage of the softness of soft manipulator, to perform minimally invasive surgery	Multi-module [67]	Fluid-driven mode
		Multibending soft robot [17] Tendon-driven continuum robot [90]	Cable-driven mode
		Soft probe [89]	Cable-driven mode
		Ferromagnetic soft robot [23]	Electromagnetic-driven mode
		Magnetically controlled soft microrobot [26]	Electromagnetic-driven mode

(continued)

Table 1.3 (continued)

Application field	Application scenario	Type of soft manipulator	Driven mode
Assist maintenance	Navigating and machining within confined workspaces to finish the operation of high value-added products (e.g. aero-engines)	Slender continuum robot [12]	Cable-driven mode
		Multi-segment continuum manipulator [88]	Cable-driven mode
Man-machine interaction	Intrinsically safe robots that are suitable for providing assistance to humans	PneuArm [8]	Fluid-driven mode
		BHA [30]	Fluid-driven mode
		HPN [29]	Fluid-driven mode
Space application	Potential in space application because of their inherent compliance, light weight, high energy densities and actuation strains	DEMES [117] Soft manipulator driven by dielectric elastomers [124]	Electroactive polymer-driven mode

lator. There are also researchers trying to adopt the shape memory materials driven mode in the underwater environment. Currently, the medical operation is the most successfully application field for the soft manipulator. In this field, the fluid-driven mode, the cable-driven mode and the electromagnetic-driven mode are all studied by the researchers. In the application field of the assist maintenance, the cable-driven mode is the popular choice for the soft manipulator. In the application field of the man-machine interaction, the fluid-driven mode, especially the pneumatic actuation is the most popular choice. The understanding of human motion and gestures will be an important capability for the robots [258]. In the application field of space manipulation, the electroactive polymer driven mode is an ideal way for the soft manipulator because it is possible to be miniaturized and the electric energy is relatively sufficient and renewable in space. In the future, the opportunities and challenges of the soft manipulator may lie in the following application prospects.

(1) Soft manipulator inspired by the origami structure [259, 260]. Origami robots are inspired by the folding processes and provide a novel way to develop a wide range of soft robots. However, the actuation of the origami structure is a great challenge. Currently, the cable-driven mode, the air-driven mode, and the

electroactive polymer-driven mode have been used in the origami robot. The cable-driven and the air-driven mechanisms are independent from the origami structure, while the electroactive polymer-driven components can be integrated with the origami structure. Thus, the combination of the smart materials and the origami structure will show a wide prospect for the soft robot.

(2) Untethered soft robot [261]. Currently, the soft robot system often requires a tethered support mechanism, such as the pneumatic or the cable control hardware. These tethered mechanisms are much more complicated and heavier than the body-part of the soft robots. This kind of fact limits its application in wearable or field robotics, where the essential components must be fully integrated and embedded within their own structure. On this background, the untethered soft robot, which tries to integrate batteries, microprocessors, pumps or motors into its own body, is studied by the scientists and engineers.

(3) Biomimetic soft microrobots [262, 263]. Microrobots can be applied in complex media and environment due to its small size. Soft microrobots are especially significant in the application field of the medical operation. However, the integration of manipulation functions in mobile robotic systems of sub-millimeter size has never been an easy task. A promising alternative strategy in developing microrobots is to implement sensing, actuation and control directly in the materials, thereby mimicking biological matter.

(4) Self-healing materials used in soft robots [264, 265]. To promote the application of the soft robot in human interaction, new soft multifunctional materials are the great challenge. These materials must be capable of self-repair and healing when damaged to maintain the longevity of the host system and prevent sudden or permanent failure. Thus, the self-healing materials that enhance the durability, mechanical robustness, and longevity of soft-matter machines and electronics will be a valuable topic for the soft robot.

(5) Fabrication of soft robots using 3D and 4D printing technology [266]. Currently, the prototypes of the soft robots are mostly fabricated by mold shaping and hand-made processes, which are complicated and inefficient. Facing these challenges, 3D and 4D printing technologies provide rapid design and fabrication approaches, which have led to an era of manufacturing in which computers can control the fabrication of soft matter that has tunable mechanical, electrical and other functional properties.

(6) Soft manipulator applied in space on orbit service (OOS) missions [267–269]. With the rapid development of space on orbit service, the need for lightweight and flexible space manipulation appears gradually [267, 270, 271]. The soft manipulator will have broad opportunities in space missions due to its lightweight structure, flexible motion and foldable characteristics. Mounted on an intelligent satellite platform, space soft manipulator will show broad application prospects in performing OOS missions, such as propellant refueling, space facility manufacturing and maintenance, like the projects "Restore-L" and "Spider" [272, 273]. It can also provide help to capture and release the target spacecraft, just like the orbital express (OE) [274] and TECSAS [275].

1.5 Summary

In this investigation, the research advance and prospect of the soft manipulator worldwide were described. Firstly, the soft manipulator was classified as fluid-driven mode, cable-driven mode, electroactive polymer driven mode, shape memory material driven mode, electromagnetic-driven mode, hybrid drive mode and so on. The typical design cases of the soft manipulator under different drive modes were summarized and compared systematically. The selection of the drive mode needs to be comprehensively analyzed and evaluated according to different task types and characteristics. Different drive modes will bring different design schemes and manufacturing technology, and further bring great differences in kinematics and dynamics performance, control algorithm and implementation. In conclusion, the selection of the drive mode is very significant to the soft manipulator, which needs to be fully demonstrated in combination with specific task scenarios and requirements.

Secondly, the enabling technologies of the soft manipulator were summarized as the following five aspects: kinematic and dynamic modeling, motion control, shape detection, dynamic tactile perception, and stiffening method. In the research of modeling and control of the soft manipulator, the kinematic modeling and motion control are more common, and the dynamic modeling and control are emerging gradually. The piece-wise constant curvature method and the Cosserat geometrically exact model are the most widely studied methods to solve the kinematic problem of the soft manipulator. Restricted by the current technical level, the open-loop motion control of the soft manipulator is still attracting a lot of research interest. The dynamics closed-loop control considering the external load effect or environmental interaction still needs further development. In the research of shape detection and dynamic tactile perception, the importance of embedded detection methods and flexible tactile sensor technology will gradually be highlighted with the rapid development of the soft robot. The flexible tactile sensor can give the soft manipulator the tactile sensing function like biological tissue, to make it better interact with the working environment and further improve its intelligent degree. In the research of the stiffening method, the stiffness change of soft structure in a certain range has been realized by using specific structures or materials under certain external excitations. The stiffening method is significant to the soft manipulator because it is an effective way to solve the contradiction between the motion flexibility and the load capacity.

In conclusion, the soft manipulator is developing gradually as an important component of the soft robot. It is mainly used in manipulation and grasping tasks with a relatively clear practical application scenario. Currently, the research on the soft manipulator has initially formed a technical system, although it is still in the primary stage. With the development of the soft manipulator, its overall technology maturity will be gradually improved. Currently, the application of the soft manipulator lies mainly in the fields of the ground rescue, the underwater grabbing, the medical operation, the assist maintenance, the man-machine interaction, the space manipulation and so on. In the future, the opportunities and challenges of the soft manipulator may lie in the following application prospects, such as origami soft manipulators, untethered soft robots, biomimetic soft microrobots, self-healing soft robots, space soft manipulators and so on. All these above researches and applications indicate that the soft manipulator has already been a worldwide hot topic and will absolutely have a bright application prospect.

References

1. Billard A, Kragic D (2019) Trends and challenges in robot manipulation. Science 364(6446):1149
2. Rus D, Tolley M (2015) Design, fabrication and control of soft robots. Nature 521:467–475
3. Singh PK, Krishna CM (2014) Continuum arm robotic manipulator: a review. Univ J Mech Eng 2(6):193–198
4. Laschi C, Mazzolai B, Cianchetti M (2016) Soft robotics: technologies and systems pushing the boundaries of robot abilities. Sci Robot 1:eaah3690
5. Xu F, Wang H (2021) Soft robotics: morphology and morphology-inspired motion strategy. IEEE/CAA J Autom Sin 8(9):1500–1522
6. Kaspar C, Ravoo BJ, van der Wiel WG et al (2021) The rise of intelligent matter. Nature 594(7863):345
7. Wehner M, Truby RL, Fitzgerald DJ et al (2016) An integrated design and fabrication strategy for entirely soft, autonomous robots. Nature 536:451–466
8. Sanan S, Lynn PS, Griffith ST (2014) Pneumatic torsional actuators for inflatable robots. J Mech Robot 6(3):031003
9. Sanan S (2013) Soft inflatable robots for safe physical human interaction. Carnegie Mellon University (PhD)
10. Mcmahan W, Chitrakaran V, Csencsits M et al (2006) Field trials and testing of the OctArm continuum manipulator. IEEE international conference on robotics and automation. Orlando, FL, pp 2336–2341
11. Grissom M D, Chitrakaran V, Dienno D, et al (2006) Design and experimental testing of the OctArm soft robot manipulator. In: Proceedings of SPIE the international society for optics and photonics, Orlando, FL, vol 6230
12. Dong X, Palmer D, Axinte D et al (2019) In-situ repair/maintenance with a continuum robotic machine tool in confined space. J Manuf Process 38:313–318
13. Wang M, Palmer D, Dong X, et al (2018) Design and development of a slender dual-structure continuum robot for in-situ aeroengine repair. In: IEEE/RSJ international conference on intelligent robots and systems, Madrid, Spain, pp 5648–5653
14. Axinte D, Dong X, Palmer D et al (2017) Development of a slender continuum robotic system for on-wing inspection/repair of gas turbine engines. Robot Comput Integ Manuf 44:218–229
15. Li C, Rahn C (2002) Design of continuous backbone, cable-driven robots. J Mech Des 124:265–271
16. Giorelli M, Renda F, Calisti M et al (2015) Learning the inverse kinetics of an octopus-like manipulator in three-dimensional space. Bioinsp Biomimet 10(3):035006
17. Renda F, Giorelli M, Calisti M et al (2014) Dynamic model of a multibending soft robot arm driven by cables. IEEE Trans Robot 30(5):1109–1122
18. Kovacs G, Düring L, Michel S et al (2009) Stacked dielectric elastomer actuator for tensile force transmission. Sens Actuat A 155:299–307
19. Qiu Y, Zhang E, Plamthottam R et al (2018) Dielectric elastomer artificial muscle: materials innovations and device explorations. Acc Chem Res 52(2):316–325
20. Pei Q, Rosenthal M, Stanford S et al (2004) Multiple-degrees-of-freedomelectroelastomer roll actuators. Smart Mater Struct 13:86–92
21. Cianchetti M, Licofonte A, Follador M et al (2014) Bioinspired soft actuation system using shape memory alloys. Actuators 3(3):226–244

22. Mazzolai B, Margheri L, Cianchetti M, et al (2012) Soft-robotic arm inspired by the octopus: II. From artificial requirements to innovative technological solutions. Bioinsp Biomimet 7(2):025005
23. Kim Y, Parada G A, Liu S, et al (2019) Ferromagnetic soft continuum robots. Science Robotics 4(33): eaax7329
24. Kim Y, Yuk H, Zhao R et al (2018) Printing ferromagnetic domains for untethered fast-transforming soft materials. Nature 558(7709):274–279
25. Liu X, Tang TC, Tham E et al (2017) Stretchable living materials and devices with hydrogel-elastomer hybrids hosting programmed cells. Proc Natl Acad Sci 114(9):2200–2205
26. Jeon S, Hoshiar AK, Kim K et al (2019) A magnetically controlled soft microrobot steering a guidewire in a three-dimensional phantom vascular network. Soft Robot 6(1):54–68
27. Lee S, Lee S, Kim S et al (2018) Fabrication and characterization of a magnetic drilling actuator for navigation in a three-dimensional phantom vascular network. Sci Rep 8(1):3691
28. Yan J, Shi P, Zhang X et al (2018) Review of biomimetic mechanism, actuation, modeling and control in soft manipulators. J Mech Eng 54(15):1–14
29. Jiang H, Liu X, Chen X et al (2016) Design and simulation analysis of a soft manipulator based on honeycomb pneumatic networks. IEEE international conference on robotics and biomimetics. Qingdao, China, pp 350–356
30. Mahl T, Hildebrandt A, Sawodny O (2014) A variable curvature continuum kinematics for kinematic control of the bionic handling assistant. IEEE Trans Robot 30(4):935–949
31. Gong Z, Fang X, Chen X et al (2021) A soft manipulator for efficient delicate grasping in shallow water: modeling, control, and real-world experiments. Int J Robot Res 40(1):449
32. Cianchetti M, Ranzani T, Gerboni G et al (2014) Soft robotics technologies to address shortcomings in today's minimally invasive surgery: the STIFF-FLOP approach. Soft Robot 1(2):122–131
33. Marchese AD, Katzschmann RK, Rus DJSR (2015) A recipe for soft fluidic elastomer robots. Soft Robot 2(1):7
34. Hawkes EW, Blumenschein LH, Greer JD et al (2017) A soft robot that navigates its environment through growth. Sci Robot 2(8):eaan3028
35. Dong H (2016) Research on key technologies of a pneumatic soft omnidirectional bending module. Harbin Institute of Technology (Master Degree)
36. Kim W, Byun J, Kim JK et al (2019) Bioinspired dual-morphing stretchable origami. Sci Robot 4(36):eaay3493
37. Guan Q, Sun J, Liu Y et al (2020) Novel bending and helical extensile/contractile pneumatic artificial muscles inspired by elephant trunk. Soft Robot 7(5):597
38. Liu X, Zhao Y, Geng D et al (2021) Soft humanoid hands with large grasping force enabled by flexible hybrid pneumatic actuators. Soft Robot 8(2):175
39. Fu HC, Ho JDL, Lee KH et al (2020) Interfacing soft and hard: a spring reinforced actuator. Soft Robot 7(1):44
40. Wang Y, Xu Q (2021) Design and testing of a soft parallel robot based on pneumatic artificial muscles for wrist rehabilitation. Sci Rep 11(1):1273
41. Zatopa A, Walker S, Menguc Y (2018) Fully soft 3D-printed electroactive fluidic valve for soft hydraulic robots. Soft Robot 5(3):258
42. Drotman D, Jadhav S, Sharp D et al (2021) Electronics-free pneumatic circuits for controlling soft-legged robots. Sci Robot 6(51):eaay2627
43. Best CM, Rupert L, Killpack MD (2021) Comparing model-based control methods for simultaneous stiffness and position control of inflatable soft robots. Int J Robot Res 40(1):470
44. Huang W, Xiao J, Xu Z (2020) A variable structure pneumatic soft robot. Sci Rep 10(1):18778
45. Teeple CB, Koutros TN, Graule MA et al (2020) Multi-segment soft robotic fingers enable robust precision grasping. Int J Robot Res 39(14):1647
46. Shen Z, Zhong H, Xu E et al (2020) An underwater robotic manipulator with soft bladders and compact depth-independent actuation. Soft Robot 7(5):535
47. Sinatra NR, Teeple CB, Vogt DM et al (2019) Ultragentle manipulation of delicate structures using a soft robotic gripper. Sci Robot 4(33):eaax5425

48. Fei Y, Wang J, Pang W (2019) A novel fabric-based versatile and stiffness-tunable soft gripper integrating soft pneumatic fingers and wrist. Soft Robot 6(1):1
49. Tawk C, In Het Panhuis M, Spinks GM et al (2018) Bioinspired 3D printable soft vacuum actuators for locomotion robots, grippers and artificial muscles. Soft Robot 5(6):685
50. Konishi S, Hirata A (2019) Flexible temperature sensor integrated with soft pneumatic microactuators for functional microfingers. Sci Rep 9(1):15634
51. Kang RJ, Branson DT, Zheng TJ et al (2013) Design, modeling and control of a pneumatically actuated manipulator inspired by biological continuum structures. Bioinsp Biomimet 8(3):036008
52. Sanan S, Ornstein MH, Atkeson CG (2011) Physical human interaction for an inflatable manipulator. In: Annual international conference of the IEEE engineering in medicine and biology society, pp 7401–7404
53. Sanan S, Moidel JB, Atkeson CG (2009) Robots with inflatable links. In: IEEE/RSJ international conference on intelligent robots and systems, pp 4331–4336
54. Jones BA, Walker ID (2006) Practical kinematics for real-time implementation of continuum robots. IEEE Trans Robot 22(6):1087–1099
55. Scott GP, Henshaw CG, Walker ID, et al (2015) Autonomous robotic refueling of an unmanned surface vehicle in varying sea states. In: IEEE/RSJ International Conference on Intelligent Robots and Systems: 1664-1671
56. Trivedi D, Lotfi A, Rahn CD (2008) Geometrically exact models for soft robotic manipulators. IEEE Trans Robot 24(4):773–780
57. Trivedi D, Lotfi A, Rahn CD (2007) Geometrically exact dynamic models for soft robotic manipulators. IEEE/RSJ International Conference on Intelligent Robots and Systems. San Diego, CA, USA, pp 1497–1502
58. Trivedi D, Dienno D, Rahn CD (2007) Optimal, model-based design of soft robotic manipulators. J Mech Des 130(9):801
59. Jones BA, Walker ID (2006) Practical kinematics for real-time implementation of continuum robots. IEEE Trans Robot 22(6):1087–1099
60. You X, Zhang Y, Chen X et al (2017) Model-free control for soft manipulators based on reinforcement Learning. In: Vancouver BC (ed) IEEE/RSJ international conference on intelligent robots and systems. Canada, pp 2909–2915
61. Jiang H, Wang Z, Liu X et al (2017) A two-level approach for solving the inverse kinematics of an extensible soft arm considering viscoelastic behavior. In: IEEE international conference on robotics and automation, Singapore, pp 6127–6133
62. Jiang H, Wang Z, Jin Y et al (2021) Hierarchical control of soft manipulators towards unstructured interactions. Int J Robot Res 40(1):411–434
63. Rolf M, Steil JJ (2013) Efficient exploratory learning of inverse kinematics on a bionic elephant trunk. IEEE Trans Neural Netw Learn Syst 25(6):1147–1160
64. Escande C, Chettibi T, Merzouki R (2014) Kinematic calibration of a multisection bionic manipulator. IEEE/ASME Trans Mechatron 20(2):663–674
65. Rolf M, Steil JJ, Gienger M (2010) Goal babbling permits direct learning of inverse kinematics. IEEE Trans Auton Ment Dev 2(3):216–229
66. Rolf M, Steil JJ (2012) Constant curvature continuum kinematics as fast approximate model for the bionic handling assistant. IEEE/RSJ international conference on intelligent robots and systems. Vilamoura-Algarve, Portugal, pp 3440–3446
67. Falco ID, Cianchetti M, Menciassi A (2017) A soft multi-module manipulator with variable stiffness for minimally invasive surgery. Bioinsp Biomimet 10(3):056008
68. Arezzo A, Mintz Y, Allaix ME et al (2017) Total mesorectal excision using a soft and flexible robotic arm: a feasibility study in cadaver models. Surg Endosc 31:264–273
69. Ranzani T, Gerboni G, Cianchetti M et al (2015) A bioinspired soft manipulator for minimally invasive surgery. Bioinsp Biomimet 10(3):035008
70. Ranzani T, Cianchetti M, Gerboni G et al (2016) A soft modular manipulator for minimally invasive surgery: design and characterization of a single module. IEEE Trans Robot 32(1):187–200

71. Thuruthel T G, Shih B, Laschi C et al (2019) Soft robot perception using embedded soft sensors and recurrent neural networks. Sci Robot 4(26):eaav1488
72. Marchese AD, Tedrake R, Rus D et al (2015) Dynamics and trajectory optimization for a soft spatial fluidic elastomer manipulator. Int J Robot Res 35(8):1000–1009
73. Marchese AD, Komorowski K et al (2014) Design and control of a soft and continuously deformable 2D robotic manipulation system. IEEE international conference on robotics and automation. China, Hong Kong, pp 2189–2196
74. Marchese AD, Katzschmann RK, Rus D (2014) Whole arm planning for a soft and highly compliant 2D robotic manipulator. IEEE/RSJ international conference on intelligent robots and systems. USA, Chicago, IL, pp 554–560
75. Marchese AD, Rus D (2015) Design, kinematics, and control of a soft spatial fluidic elastomer manipulator. Int J Robot Res 35(7):840–869
76. Katzschmann RK, Marchese AD, Rus D (2015) Autonomous object manipulation using a soft planar grasping manipulator. Soft Robot 2(4):155–165
77. Greer JD, Morimoto TK, Okamura AM et al (2016) A soft, steerable continuum robot that grows via tip extension. Soft Robot 6(1):95–108
78. Greer JD, Blumenschein LH, Okamura AM et al (2018) Obstacle-aided navigation of a soft growing robot. IEEE international conference on robotics and automation. Brisbane, Australia, pp 4165–4172
79. El-Hussieny H, Mehmood U, Mehdi Z et al (2018) Development and evaluation of an intuitive flexible interface for teleoperating soft growing robots. In: IEEE/RSJ international conference on intelligent robots and systems, Madrid, Spain, pp 4995–5002
80. Hammond ZM, Usevitch NS, Hawkes EW et al (2017) Pneumatic reel actuator: design, modeling, and implementation. In: IEEE international conference on robotics and automation, Singapore, pp 626–633
81. Greer JD, Morimoto TK, Okamura AM et al (2017) Series pneumatic artificial muscles (sPAMs) and application to a soft continuum robot. In: IEEE international conference on robotics and automation, Singapore, pp 5503–5510
82. Yan J, Xu B, Zhang X et al (2017) Design and test of a new spiral driven pure torsional soft actuator. Springer International Publishing, Cham
83. Yan J, Zhang X, Xu B et al (2018) A new spiral-type inflatable pure torsional soft actuator. Soft Robot 5(5):527
84. Truby RL, Katzschmann RK, Lewis JA et al (2019) Soft robotic fingers with embedded lonogel sensors and discrete actuation modes for somatosensitivemanipulation. The 2nd IEEE international conference on soft robotics. Seoul, Korea, pp 322–329
85. Truby RL, Wehner M, Grosskopf AK et al (2018) Soft somatosensitive actuators via embedded 3D printing. Adv Mater 30(15):1706383
86. Garriga-Casanovas A, Collison I, Rodriguez Y (2018) Toward a common rramework for the design of soft robotic manipulators with fluidic actuation. Soft Robot 5(5):622
87. Palacio JMÀ, Riwan A, Mechbal N et al (2017) A novel inflatable actuator for inflatable robotic arms. IEEE international conference on advanced intelligent mechatronics. Munich, Germany, pp 88–93
88. Yu X, Wang X, Meng D et al (2018) Collison free path planning for multi-section continuum manipulators based on a model method. IEEE 8th annual international conference on CYBER technology in automation, control, and intelligent systems. Tianjin, China, pp 236–242
89. Deng T, Wang H, Chen W et al (2013) Development of a new cable-driven soft robot for cardiac ablation. IEEE international conference on robotics and biomimetics. Shenzhen, China, pp 728–733
90. Amanov E, Nguyen TD, Burgner-Kahrs J (2021) Tendon-driven continuum robots with extensible sections–a model-based evaluation of path-following motions. Int J Robot Res 40(1):7
91. Hussain I, Al-Ketan O, Renda F et al (2020) Design and prototyping soft-rigid tendon-driven modular grippers using interpenetrating phase composites materials. Int J Robot Res 39(14):1635

92. Mutlu R, Alici G, In Het Panhuis M et al (2016) 3D printed flexure hinges for soft monolithic prosthetic fingers. Soft Robot 3(3):120
93. Morales BT, Kruszewski A, Carrez B et al (2020) Design, implementation, and control of a deformable manipulator robot based on a compliant spine. Int J Robot Res 39(14):1604
94. Alatorre D, Nasser B, Rabani A et al (2018) Teleoperated, in situ repair of an aeroengine: overcoming the internet latency hurdle. IEEE Robot Autom Mag 26(1):10–20
95. Dong X, Raffles M, Cobos-Guzman S et al (2015) A novel continuum robot using twin-pivot compliant joints: design, modeling, and validation. J Mech Robot 8(2):021010
96. Barrientos-Diez J, Dong X, Axinte D et al (2021) Real-time kinematics of continuum robots: modelling and validation. Robot Comput Integ Manuf 67:102019
97. Dong X, Raffles M, Guzman SC et al (2014) Design and analysis of a family of snake arm robots connected by compliant joints. Mech Mach Theory 77:73–91
98. Giorgio-Serchi F, Arienti A, Corucci F et al (2017) Hybrid parameter identification of a multi-modal underwater soft robot. Bioinsp Biomimet 12:025007
99. Giorelli M, Renda F, Calisti M et al (2015) Neural network and jacobian method for solving the inverse statics of a cable-driven soft arm with nonconstant curvature. IEEE Trans Robot 31(4):823–834
100. Renda F, Cianchetti M, Giorelli M et al (2012) A 3D steady-state model of a tendon-driven continuum soft manipulator inspired by the octopus arm. Bioinsp Biomimet 7(2):025006
101. Michele G, Federico R, Gabriele et al (2013) A feed-forward neural network learning the inverse kinetics of a soft cable-driven manipulator moving in three-dimensional space. IEEE/RSJ international conference on intelligent robots and systems. Tokyo, Japan, pp 5033–5039
102. Voisembert S, Mechbal N, Riwan A et al (2013) Design of a novel long-range inflatable robotic arm: manufacturing and numerical evaluation of the joints and actuation. J Mech Robot 5(4):045001
103. Voisembert S, Riwan A, Mechbal N (2012) Numerical evaluation of a new robotic manipulator based on inflatable joints. IEEE international conference on automation science and engineering. Seoul, Korea (South), pp 544–549
104. Voisembert S, Riwan A, Mechbal N et al (2011) A novel inflatable robot with constant and continuous volume. IEEE international conference on robotics and automation. Shanghai, China, pp 5843–5848
105. Voisembert S, Mechbal N, Riwan A et al (2011) A novel inflatable tendon driven manipulator with constant volume. International design engineering technical conferences and computers and information in engineering conference. DC, USA, Washington, pp 1233–1242
106. Riwan A, Voisembert S (2013) Articulated inflatable structure and robot arm comprising such a structure. U.S. Patent Application 13/700,945
107. Zhang Z, Wang X, Wang S et al (2018) Shape detection and reconstruction of soft robotic arm based on fiber bragg grating sensor array. IEEE international conference on robotics and biomimetics. Kuala Lumpur, Malaysia, pp 978–983
108. Huang S, Meng D, She Y et al (2018) Statics of continuum space manipulators with nonconstant curvature via pseudorigid-body 3R model. IEEE Acces 6:70854–70865
109. Yu X, Wang X, Meng D et al (2018) Differential kinematics for a tendon-driven snake-like robot. Chin Control Decis Conf. Shenyang, China, pp 5768–5773
110. Xu F, Wang H, Au KWS et al (2018) Underwater dynamic modeling for a cable-driven soft robot arm. IEEE/ASME Trans Mechatron 23(6):2726–2738
111. Wang H, Zhang R, Chen W et al (2016) Shape detection algorithm for soft manipulator based on fiber bragg gratings. IEEE/ASME Trans Mechatron 21(6):2977–2982
112. Wang H, Wang C, Chen W et al (2016) Three-dimensional dynamics for cable-driven soft manipulator. IEEE/ASME Trans Mechatron 22(1):18–28
113. Zhang R, Wang H, Chen W et al (2016) Shape control for a soft robot inspired by octopus. Robot 38(6):754
114. Wang H, Yang B, Liu Y et al (2016) Visual servoing of soft robot manipulator in constrained environments with an adaptive controller. IEEE/ASME Trans Mechatron 22(1):41–50

115. Luo Y, Na Z, Wang H et al (2017) Design, modeling and experimental validation of a scissor mechanisms enabled compliant modular earthworm-like robot. In: Vancouver BC (ed) IEEE/RSJ international conference on intelligent robots and systems. Canada, pp 2421–2426
116. Hussain I, Malvezzi M, Gan D et al (2021) Compliant gripper design, prototyping, and modeling using screw theory formulation. Int J Robot Res 40(1):55–71
117. Araromi OA, Gavrilovich I, Shintake J et al (2014) Rollable multisegment dielectric elastomer minimum energy structures for a deployable microsatellite gripper. IEEE/ASME Trans Mechatron 20(1):438–446
118. Choi HR, Jung K, Ryew S et al (2005) Biomimetic soft actuator: Design, modeling, control, and applications. IEEE/ASME Trans Mechatron 10(5):581–593
119. Jung K, Koo JC, Lee YK et al (2007) Artificial annelid robot driven by soft actuators. Bioinsp Biomimet 2(2):S42
120. Miriyev A, Stack K, Lipson H (2017) Soft material for soft actuators. Nat Commun 8(1):596
121. Zhao Y, Lo C Y, Ruan L, et al (2021) Somatosensory actuator based on stretchable conductive photothermally responsive hydrogel. Sci Robot 6(53):eabd5483
122. Lee H, Jang Y, Choe JK et al (2020) 3D-printed programmable tensegrity for soft robotics. Sci Robot 5(45):eaay9024
123. He Q, Wang Z, Wang Y, et al (2019) Electrically controlled liquid crystal elastomer-based soft tubular actuator with multimodal actuation. Sci Adv 5(10):eaax5746
124. Xing Z, Zhang J, McCoul D et al (2020) A super-lightweight and soft manipulator driven by dielectric elastomers. Soft Robot 7(4):512–520
125. Araromi OA, Rosset S, Shea HR (2015) Versatile fabrication of PDMS-carbon electrodes for silicone dielectric elastomer transducers. International conference on solid-state sensors. Actuators and Microsystems, Anchorage, AK, pp 1905–1908
126. Rosset S, Araromi OA, Shintake J et al (2014) Model and design of dielectric elastomer minimum energy structures. Smart Mater Struct 23(8):085021
127. Araromi OA, Conn AT, Ling CS et al (2011) Spray deposited multilayered dielectric elastomer actuators. Sens Actuat A 167(2):459–467
128. Pei Q, Pelrine R, Rosenthal MA et al (2004) Recent progress on electroelastomer artificial muscles and their application for biomimetic robots. In: Smart structures and materials 2004: electroactive polymer actuators and devices, vol 5385, pp 41–50
129. Pei Q, Pelrine R, Stanford S et al (2003) Electroelastomer rolls and their application for biomimetic walking robots. Synth Met 135:129–131
130. Nguyen CT, Phung H, Nguyen TD et al (2014) A small biomimetic quadruped robot driven by multistacked dielectric elastomer actuators. Smart Mater Struct 23:065005
131. Jung HS, Kim Y, Hoang T et al (2018) Musclelike joint mechanism driven by dielectric elastomer actuator for robotic applications. Smart Mater Struct 27(7):075011
132. Lu T, Huang J, Jordi C et al (2012) Dielectric elastomer actuators under equal-biaxial forces, uniaxial forces, and uniaxial constraint of stiff fibers. Soft Matter 8(22):6167–6173
133. Weiss FM, Töpper T, Osmani B et al (2016) Electrospraying nanometer-thin elastomer films for low voltage dielectric actuators. Adv Electron Mater 2(5):1500476
134. Cheng SS, Kim Y, Desai JP (2017) New actuation mechanism for actively cooled SMA springs in a neurosurgical robot. IEEE Trans Robot 33(4):986–993
135. Zhang L, Xu M, Yang H (2017) Research on soft manipulator actuated by shape memory alloy (SMA) springs. IEEE international conference on real-time computing and robotics. Okinawa, Japan, pp 74–78
136. Wang W, Yu CY, Abrego Serrano PA et al (2020) Shape memory alloy-based soft finger with changeable bending length using targeted variable stiffness. Soft Robot 7(3):283–291
137. Wang W, Ahn SH (2017) Shape memory alloy-based soft gripper with variable stiffness for compliant and effective grasping. Soft Robot 4(4):379–389
138. Margheri L, Laschi C, Mazzolai, B (2012) Soft robotic arm inspired by the octopus: I. From biological functions to artificial requirements. Bioinsp Biomimet 7(2):025004
139. Laschi C, Cianchetti M, Mazzolai B et al (2012) Soft robot arm inspired by the octopus. Adv Robot 26(7):709–727

140. Cianchetti M, Arienti A, Follador M et al (2011) Design concept and validation of a robotic arm inspired by the octopus. Mater Sci Eng, C 31(6):1230–1239
141. Yang H, Xu M, Li W et al (2018) Design and implementation of a soft robotic arm driven by SMA coils. IEEE Trans Industr Electron 66(8):6108–6116
142. Zhang L, Xu M, Yang H et al (2017) Research on flexible manipulator driven by shape memory alloy. Mach Electron 35(6):72–76
143. Kim J, Chung SE, Choi SE et al (2011) Programming magnetic anisotropy in polymeric microactuators. Nat Mater 10(10):747
144. Wang X, Mao GY, Ge J et al (2011) Untethered and ultrafast soft-bodied robots. Commun Mater 1(1):1
145. Mao G, Drack M, Karami-Mosammam M, et al (2020) Soft electromagnetic actuators. Sci Adv 6(26):eabc0251
146. Xu T, Zhang J, Salehizadeh M et al (2019) Millimeter-scale flexible robots with programmable three-dimensional magnetization and motions. Sci Robot 4(29):eaav4494
147. Jang M, Kim JS, Kim JH et al (2019) Surface-controlled molecular self-alignment in polymer actuators for flexible microrobot applications. Polymers 11(4):736
148. Jeon S, Hoshiar AK, Kim S et al (2018) Improving guidewire-mediated steerability of a magnetically actuated flexible microrobot. Micro Nano Syst Lett 6(15):1–10
149. Hoshiar AK, Jeon S, Kim K et al (2018) Steering algorithm for a flexible microrobot to enhance guidewire control in a coronary angioplasty application. Micromachines 9(617):1–13
150. Manti M, Pratesi A, Falotico E et al (2016) Soft assistive robot for personal care of elderly people. In: IEEE international conference on biomedical robotics and biomechatronics, Singapore, pp 833–838
151. Yang C, Geng S, Walker I et al (2020) Geometric constraint-based modeling and analysis of a novel continuum robot with shape memory alloy initiated variable stiffness. Int J Robot Res 39(14):1620
152. Diteesawat RS, Helps T, Taghavi M et al (2021) Electro-pneumatic pumps for soft robotics. Sci Robot 6(51):eabc3721
153. Morimoto Y, Onoe H, Takeuchi S et al (2018) Biohybrid robot powered by an antagonistic pair of skeletal muscle tissues. Sci Robot 3(18):eaat4440
154. Acome E, Mitchell SK, Morrissey TG et al (2018) Hydraulically amplified self-healing electrostatic actuators with muscle-like performance. Science 359(6371):61–65
155. Subramaniam V, Jain S, Agarwal J et al (2020) Design and characterization of a hybrid soft gripper with active palm pose control. Int J Robot Res 39(14):1668
156. Tang Y, Chi Y, Sun J, et al (2020) Leveraging elastic instabilities for amplified performance: Spine-inspired high-speed and high-force soft robots. Sci Adv 6(19):eaaz6912
157. Li C, Lau G C, Yuan H, et al (2020) Fast and programmable locomotion of hydrogel-metal hybrids under light and magnetic fields. Sci Robot 5(49):eabb9822
158. Manti M, Thuruthel TG, Falotico FP et al (2017) Exploiting morphology of a soft manipulator for assistive tasks. Conference on biomimetic and biohybrid systems. CA, USA, Stanford, pp 291–301
159. Renda F, Cianchetti M, Abidi H et al (2017) Screw-based modeling of soft manipulators with tendon and fluidic actuation. J Mech Robot 9(4):040012
160. Kang R, Guo Y, Chen L et al (2016) Design of a pneumatic muscle based continuum robot with embedded tendons. IEEE/ASME Trans Mechatron 22(2):751–761
161. Chen X, Zhang X, Liu H et al (2020) Design and development of a soft robotic manipulator. Int J Mech Mater Des 16:309–321
162. Liu Z, Zhang X, Liu H et al (2020) Kinematic modelling and experimental validation of a foldable pneumatic soft manipulator. Appl Sci 10(4):1447
163. Jones BA, Walker ID (2006) Kinematics for multisection continuum robots. IEEE Trans Robot 22(1):43–55
164. Hannan MW, Walker ID (2003) Kinematics and the implementation of an elephant's trunk manipulator and other continuum style robots. J Robot Syst 20(2):45–63

165. Habibi H, Yang C, Kang R et al (2018) Modelling an actuated large deformation soft continuum robot surface undergoing external forces using a lumped-mass approach. In: IEEE/RSJ international conference on intelligent robots and systems, Madrid, Spain, pp 5958–5963
166. Mengacci R, Angelini F, Catalano MG et al (2021) On the motion/stiffness decoupling property of articulated soft robots with application to model-free torque iterative learning control. Int J Robot Res 40(1):348
167. Lee KH, Fu DK, Leong MC et al (2017) Nonparametric online learning control for soft continuum robot: an enabling technique for effective endoscopic navigation. Soft Robot 4(4):324–337
168. Alici G, Canty T, Mutlu R et al (2018) Modeling and experimental evaluation of bending behavior of soft pneumatic actuators made of discrete actuation chambers. Soft Robot 5(1):24–35
169. Xu P A, Mishra A K, Bai H, et al (2019) Optical lace for synthetic afferent neural networks. Sci Robot 4(34):eaaw6304
170. Galloway K C, Chen Y, Templeton E et al. Fiber optic shape sensing for soft robotics. Soft Robot 6(5):671
171. Kim S Y, Choo Y, Bilodeau RA et al (2020) Sustainable manufacturing of sensors onto soft systems using self-coagulating conductive pickering emulsions. Sci Robot 5(39):eaay3604
172. Guo J, Elgeneidy K, Xiang C et al (2018) Soft pneumatic grippers embedded with stretchable electroadhesion. Smart Mater Struct 27(5):055006
173. Chen Y, Guo S, Li C et al (2018) Size recognition and adaptive grasping using an integration of actuating and sensing soft pneumatic gripper. Robot Auton Syst 104:14–24
174. Yan Y, Hu Z, Yang Z et al (2021) Soft magnetic skin for super-resolution tactile sensing with force self-decoupling. Sci Robot 6(51):eabc8801
175. Justus KB, Hellebrekers T, Lewis DD et al (2019) A biosensing soft robot: autonomous parsing of chemical signals through integrated organic and inorganic interfaces. Sci Robot 4(31):eaax0765
176. Sankar S, Balamurugan D, Brown A et al (2021) Texture discrimination with a soft biomimetic finger using a flexible neuromorphic tactile sensor array that provides sensory feedback. Soft Robot 8(5):577–587
177. Hao Y, Wang T, Xi F et al (2017) A variable stiffness soft robotic gripper with low-melting-point alloy. Chinese control conference. Dali, China, pp 6781–6786
178. Hao Y, Wang T, Xie Z et al (2018) A eutectic-alloy-infused soft actuator with sensing, tunable degrees of freedom, and stiffness properties. J Micromech Microeng 28(2):024004
179. Shintake J, Schubert B, Rosset S et al (2015) Variable stiffness actuator for soft robotics using dielectric elastomer and low-melting-point alloy. In: IEEE/RSJ international conference on intelligent robots and systems. Hamburg, Germany, pp 1097–1102
180. Naselli GA, Mazzolai B (2021) The softness distribution index: towards the creation of guidelines for the modeling of soft-bodied robots. Int J Robot Res 40(1):197
181. Camarillo DB, Milne CF, Carlson CR et al (2008) Mechanics modeling of tendon-driven continuum manipulators. IEEE Trans Robot 24(6):1262–1273
182. Webster RJ, Jones BA (2010) Design and kinematic modeling of constant curvature continuum robots: a review. Int J Robot Res 29(13):1661–1683
183. Camarillo DB, Carlson CR, Salisbury JK (2009) Configuration tracking for continuum manipulators with coupled tendon drive. IEEE Trans Robot 25(4):798–808
184. Chirikjian GS (1993) A continuum approach to hyper-redundant manipulator dynamics. In: IEEE/RSJ international conference on intelligent robots and systems Yokohama Japan, pp 1059–1066
185. Chirikjian GS (1995) Hyper-redundant manipulator dynamics: a continuumapproximation. Adv Robot 9(3):217–243
186. Chirikjian GS, Burdick JW (1995) The kinematics of hyper-redundant robot locomotion. IEEE Trans Robot Autom 11(6):781–794
187. Grazioso S, di Gironimo G, Siciliano B (2019) A geometrically exact model for soft continuum robots: the finite element deformation space formulation. Soft Robot 6(6):790

188. Renda F, Laschi C (2012) A general mechanical model for tendon-driven continuum manipulators. IEEE international conference on robotics and automation. Saint Paul, MN, USA, pp 3813–3818
189. Rucker DC, Webster RJ (2011) Statics and dynamics of continuum robots with general tendon routing and external loading. IEEE Trans Robot 27(6):1033–1044
190. Rucker DC, Jones BA, Webster RJ (2010) A geometrically exact model for externally loaded concentric-tube continuum robots. IEEE Trans Robot 26(5):769–780
191. Olson G, Chow S, Nicolai A et al (2021) A generalizable equilibrium model for bending soft arms with longitudinal actuators. Int J Robot Res 40(1):148–177
192. Tutcu C, Baydere BA, Talas SK et al (2021) Quasi-static modeling of a novel growing soft-continuum robot. Int J Robot Res 40(1):86–98
193. Shiva A, Sadati SMH, Noh Y et al (2019) Elasticity versus hyperelasticity considerations in quasistatic modeling of a soft finger-like robotic appendage for real-time position and force estimation. Soft Robot 6(2):228
194. Shan X, Birglen L (2020) Modeling and analysis of soft robotic fingers using the fin ray effect. Int J Robot Res 39(14):1686–1705
195. Sedal A, Wineman A, Gillespie RB et al (2021) Comparison and experimental validation of predictive models for soft, fiber-reinforced actuators. Int J Robot Res 40(1):119–135
196. Jeong U, Kim K, Kim SH et al (2021) Reliability analysis of a tendon-driven actuation for soft robots. Int J Robot Res 40(1):494–511
197. Hamaya M, Matsubara T, Teramae T et al (2021) Design of physical user-robot interactions for model identification of soft actuators on exoskeleton robots. Int J Robot Res 40(1):397–410
198. Singh I, Amara Y, Melingui A et al (2018) Modeling of continuum manipulators using pythagorean hodograph curves. Soft Robot 5(4):425–442
199. Sadati SMH, Naghibi SE, Shiva A et al (2021) TMTDyn: a MATLAB package for modeling and control of hybrid rigid-continuum robots based on discretized lumped systems and reduced-order models. Int J Robot Res 40(1):296–347
200. Camp AS, Chapman EM, Jaramillo Cienfuegos P (2021) Modeling and analysis of hydraulic piston actuation of McKibben fluidic artificial muscles for hand rehabilitation. Int J Robot Res 40(1):136–147
201. Digumarti KM, Trimmer B, Conn AT et al (2019) Quantifying dynamic shapes in soft morphologies. Soft Robot 6(6):733–744
202. Luo M, Agheli M, Onal CD (2014) Theoretical modeling and experimental analysis of a pressure-operated soft robotic snake. Soft Robot 1(2):136–146
203. Mustaza SM, Elsayed Y, Lekakou C et al. Dynamic modeling of fiber-reinforced soft manipulator: a visco-hyperelastic material-based continuum mechanics approach. Soft Robot 6(3):305–317
204. George Thuruthel T, Ansari Y, Falotico E et al (2018) Control strategies for soft robotic manipulators: a survey. Soft Robot 5(2):149–163
205. Abbasi P, Nekoui MA, Zareinejad M et al (2020) Position and force control of a soft pneumatic actuator. Soft Robot 7(5):550–563
206. Trumić M, Jovanović K, Fagiolini A (2021) Decoupled nonlinear adaptive control of position and stiffness for pneumatic soft robots. Int J Robot Res 40(1):277–295
207. Luo M, Skorina EH, Tao W et al (2017) Toward modular soft robotics: Proprioceptive curvature sensing and sliding-mode control of soft bidirectional bending modules. Soft Robot 4(2):117–125
208. Pozzi M, Marullo S, Salvietti G et al (2020) Hand closure model for planning top grasps with soft robotic hands. Int J Robot Res 39(14):1706–1723
209. Greer JD, Blumenschein LH, Alterovitz R et al (2020) Robust navigation of a soft growing robot by exploiting contact with the environment. Int J Robot Res 39(14):1724–1738
210. Tang ZQ, Heung HL, Tong KY et al (2021) Model-based online learning and adaptive control for a "human-wearable soft robotâŁž integrated system. Int J Robot Res 40(1):256–276
211. Franco E, Garriga-Casanovas A (2021) Energy-shaping control of soft continuum manipulators with in-plane disturbances. Int J Robot Res 40(1):236–255

212. Zhang X, Liu J, Li Y (2022) An obstacle avoidance algorithm for space hyper-redundant manipulators using combination of RRT and shape control method. Robotica 40(4):1036–1069
213. Yip MC, Camarillo DB (2016) Model-less hybrid position/force control: A minimalist approach for continuum manipulators in unknown, constrained environments. IEEE Robot Autom Lett 1(2):844–851
214. Yip MC, Camarillo DB (2014) Model-less feedback control of continuum manipulators in constrained environments. IEEE Trans Robot 30(4):880–889
215. Cherubini A, Ortenzi V, Cosgun A et al (2020) Model-free vision-based shaping of deformable plastic materials. Int J Robot Res 39(14):1739–1759
216. Zhang D, Yuan H, Cao Z (2020) Environmental adaptive control of a snake-like robot with variable stiffness actuators. IEEE/CAA J Autom Sin 7(3):745–751
217. Khan AH, Shao Z, Li S et al (2020) Which is the best PID variant for pneumatic soft robots? an experimental study. IEEE/CAA J Autom Sin 7(2):451
218. Li M, Kang R, Branson DT et al (2017) Model-free control for continuum robots based on an adaptive Kalman filter. IEEE/ASME Trans Mechatron 23(1):286–297
219. Camarillo DB, Carlson CR, Salisbury JK (2009) Task-space control of continuum manipulators with coupled tendon drive. In: Experimental robotics: the eleventh international symposium, Berlin, Heidelberg, pp 271–280
220. Sefati S, Alambeigi F, Iordachita I et al (2018) On the effect of vibration on shape sensing of continuum manipulators using fiber Bragg gratings. International symposium on medical robotics. Atlanta, GA, USA, pp 1–6
221. Sefati S, Alambeigi F, Iordachita I et al (2016) FBG-based large deflection shape sensing of a continuum manipulator: manufacturing optimization. IEEE sensors. Oriando, FL, USA, pp 1–3
222. Kim T, Lee S, Hong T, et al (2020) Heterogeneous sensing in a multifunctional soft sensor for human-robot interfaces. Sci Robot 5(49):eabc6878
223. van Meerbeek IM, De Sa CM, Shepherd RF (2018) Soft optoelectronic sensory foams with proprioception. Sci Robot 3(24):eaau2489
224. Cui J, Trinkle J (2021) Toward next-generation learned robot manipulation. Sci Robot 6(54):eabd9461
225. Man K, Damasio A (2019) Homeostasis and soft robotics in the design of feeling machines. Nat Mach Intell 1(10):446–452
226. Shih B, Shah D, Li J et al (2020) Electronic skins and machine learning for intelligent soft robots. Sci Robot 5(41):eaaz9239
227. Liu G, Xu K, Yu Q et al (2018) Advances in electronic skin research. Sci China Inf Sci 48(6):626–634
228. Song A (2020) Review of robot tactile sensors. Meas Control Technol 2:1–7
229. Qiu L, Cao J, Zhou J et al (2019) Research progress of flexible and elastic bionic e-skin for robot. J Central South Univ 50(5):1065–1074
230. Din S, Xu W, Cheng LK et al (2017) A stretchable multimodal sensor for soft robotic applications. IEEE Sens J 17(17):5678–5686
231. Preechayasomboon P, Rombokas E (2019) ConTact sensors: a tactile sensor readily integrable into soft robots. IEEE international conference on soft robotics. Seoul, Korea, pp 605–610
232. Suh C, Margarit JC, Song YS et al (2014) Soft pneumatic actuator skin with embedded sensors. IEEE/RSJ international conference on intelligent robots and systems. IL, USA, Chicago, pp 2783–2788
233. Yildiz SK, Mutlu R, Alici G et al (2016) Performance quantification of strain sensors for flexible manipulators. In: Banff AB (ed) IEEE international conference on advanced intelligent mechatronics. Canada, pp 584–589
234. Li G, Liu S, Wang L, et al(2020) Skin-inspired quadruple tactile sensors integrated on a robot hand enable object recognition. Sci Robot 5(49):eabc8134
235. Byun J, Lee Y, Yoon J et al (2018) Electronic skins for soft, compact, reversible assembly of wirelessly activated fully soft robots. Sci Robot 3(18):eaas9020

236. Mura D, Knoop E, Catalano MG et al (2020) On the role of stiffness and synchronization in human-robot handshaking. Int J Robot Res 39(14):1796–1811
237. Jin T, Sun Z, Li L et al (2020) Triboelectric nanogenerator sensors for soft robotics aiming at digital twin applications. Nat Commun 11(1):5381
238. Mintchev S, Salerno M, Cherpillod A et al (2019) A portable three-degrees-of-freedom force feedback origami robot for human-robot interactions. Nature Machine Intelligence 1(12):584
239. Shim H, Sim K, Ershad F et al (2019) Stretchable elastic synaptic transistors for neurologically integrated soft engineering systems. Sci Adv 5(10):eaax4961
240. Kim YJ, Cheng S, Kim S et al (2013) A novel layer jamming mechanism with tunable stiffness capability for minimally invasive surgery. IEEE Trans Robot 29(4):1031–1042
241. Mintchev S, Shintake J, Floreano D (2018) Bioinspired dual-stiffness origami. Sci Robot 3(20):eaau0275
242. Brancadoro M, Manti M, Tognarelli S et al (2020) Fiber jamming transition as a stiffening mechanism for soft robotics. Soft Robot 7(6):663–674
243. Al Abeach LAT, Nefti-Meziani S, Davis S (2017) Design of a variable stiffness soft dexterous gripper. Soft Robot 4(3):274–284
244. Laschi C, Mazzolai B, Mattoli V et al (2009) Design of a biomimetic robotic octopus arm. Bioinsp Biomimet 4(1):015006
245. Cianchetti M, Follador M, Mazzolai B et al (2012) Design and development of a soft robotic octopus arm exploiting embodied intelligence. IEEE international conference on robotics and automation. Saint Paul, Minnesota, USA, pp 5271–5276
246. Shahid Z, Glatman AL, Ryu SC (2019) Design of a soft composite finger with adjustable joint stiffness. Soft Robot 6(6):722–732
247. Manti M, Cacucciolo V, Cianchetti M (2016) Stiffening in soft robotics: A review of the state of the art. IEEE Robot Autom Mag 23(3):93–106
248. Majidi C, Wood RJ (2010) Tunable elastic stiffness with microconfined domains at low magnetic field. Appl Phys Lett 97(164104):1–4
249. Pettersson A, Davis S, Gray JO et al (2010) Design of a magnetorheological robot gripper for handling of delicate food products with varying shapes. J Food Eng 98:332–338
250. Sadeghi A, Beccai L, Mazzolai B (2012) Innovative soft robots based on electro-rheological fluids. IEEE/RSJ international conference on intelligent robots and systems. Vilamoura, Algarve, Portugal, pp 4237–4242
251. Hao Y, Wang T, Wen L (2017) A programmable mechanical freedom and variable stiffness soft actuator with low melting point alloy. In: The 10th international conference of intelligent robotics and applications, Wuhan, China, p 10462
252. Chautems C, Tonazzini A, Floreano D et al (2017) A variable stiffness catheter controlled with an external magnetic field. In: Vancouver BC (ed) IEEE/RSJ international conference on intelligent robots and systems. Canada, pp 181–186
253. Alambeigi F, Seifabadi R, Armand M (2016) A continuum manipulator with phase changing alloy. In: IEEE international conference on robotics and automation Stockholm, Sweden, pp 758–764
254. Must I, Sinibaldi E, Mazzolai B et al (2019) A variable-stiffness tendril-like soft robot based on reversible osmotic actuation. Nat Commun 10(1):344
255. Delmerico J, Mintchev S, Giusti A et al (2019) The current state and future outlook of rescue robotics. J Field Robot 36(7):1171–1191
256. Choi B, Lee W, Park G et al (2019) Development and control of a military rescue robot for casualty extraction task. J Field Robot 36(4):656–676
257. Cordie TP, Bandyopadhyay T, Roberts J et al (2019) Modular field robot deployment for inspection of dilapidated buildings. J Field Robot 36(4):641–655
258. Islam MJ, Ho M, Sattar J (2019) Understanding human motion and gestures for underwater human-robot collaboration. J Field Robot 36(5):851–873
259. Rus D, Tolley MT (2018) Design, fabrication and control of origami robots. Nat Rev Mater 3(6):101–112

260. Yang H, Yeow B S, Li Z et al (2019) Multifunctional metallic backbones for origami robotics with strain sensing and wireless communication capabilities. Sci Robot 4(33):eaax7020
261. Rich SI, Wood RJ, Majidi C (2018) Untethered soft robotics Nat Electron 1(2):102–112
262. Palagi S, Fischer P (2018) Bioinspired microrobots Nat Rev Mater 3(6):113–124
263. Shi C, Luu DK, Yang Q et al (2016) Recent advances in nanorobotic manipulation inside scanning electron microscopes. Microsyst Nanoeng 2(1):1–16
264. Yu K, Xin A, Du H et al (2019) Additive manufacturing of self-healing elastomers. NPG Asia Mater 11(1):7
265. Bartlett MD, Dickey MD, Majidi C (2019) Self-healing materials for soft-matter machines and electronics. NPG Asia Mater 11(1):21
266. Truby RL, Lewis JA (2016) Printing soft matter in three dimensions. Nature 540(7633):371–378
267. Jing Z, Qiao L, Pan H et al (2017) An overview of the configuration and manipulation of soft robotics for on-orbit servicing. Sci China Inf Sci 60(5):050201
268. Flores-Abad A, Ma O, Pham K et al (2014) A review of space robotics technologies for on-orbit servicing. Prog Aerosp Sci 68:1–26
269. Lin HT, Leisk GG, Trimmer BA (2013) Soft robots in space: a perspective for soft robotics. Acta Futura 6:69–79
270. Chen X, Yuan J, Yao W et al (2009) On Orbit service technology of spacecraft. China Astronautic Publishing House, Beijing
271. Zhang X, Liu H, Liu Z et al (2019) Dynamical modeling and intelligent control of space soft manipulator. Chin J Intell Sci Technol 1(1):52–61
272. Wang X (2017) Development of foreign on-orbit service systems (II). Space Int 11(467):65–69
273. Wang X (2017) Development of foreign on-orbit service systems (I). Space Int 10(466):24–29
274. Friend RB (2008) Orbital express program summary and mission overview. In: Sensors and systems for space applications II, Oxlondo, pp 11–21
275. Sommer B (2004) Automation and robotics in the German space program-unmanned on-orbit servicing (OOS) & The TECSAS Mission. In: 55th International astronautical congress of the international astronautical federation. The International Academy of Astronautics, and The International Institute of Space Law, Vancouver, British Columbia, Canada, p IAA-3

Chapter 2
Design and Development of an Elephant Trunk-Like Soft Robotic Manipulator

In this chapter, a pneumatic soft manipulator as well as its experiment system are designed and developed. In Sect. 2.1, the design schematic and the kinematic model of the soft actuator are presented. The experiment setup used to measure the elastic stiffness, which is necessary in the numerical calculation of the kinematic model, is designed and developed. In Sect. 2.2, the kinematic model of the soft manipulator, which is an assembly of three soft actuators, is deduced. The shape of the soft manipulator changes with the increase of the gas pressure and is calculated and analyzed in the numerical examples. In Sect. 2.3, the prototype of the soft manipulator is developed by stacking three soft actuators, which are printed using the selective laser sintering (SLS) technology. The pneumatic control system and the position measurement system are designed and developed. The correctness of the proposed model and the end repeat positioning accuracy are validated experimentally. Finally, the conclusions are presented.

2.1 Design of a Pneumatic Soft Actuator

2.1.1 Design Schematic

The pneumatic soft actuator is the basic module of the soft manipulator [1, 2]. Three pneumatic soft actuators are stacked in series to form the pneumatic soft manipulator. A schematic of the design of the pneumatic soft actuator is shown in Fig. 2.1a. The soft actuator is made of three inflatable bellows, which are connected and fixed by a series of thin rigid plates. The bellows will elongate in response to the pressured gas. The movement and constraints between the bellows and the thin rigid plates lead to the constrained orientation of the soft actuator. The extending and bending direction of the actuator under the pressured gas is shown in Fig. 2.1b.

X. Zhang et al., *Design and Implementation of Soft Robotic Manipulators*,
https://doi.org/10.1007/978-981-95-1657-5_2

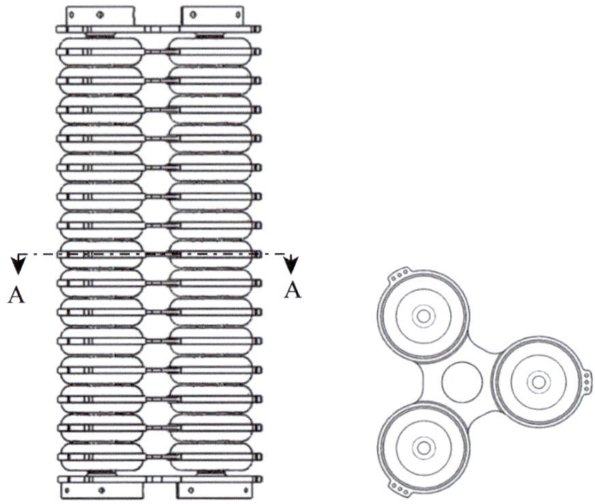

(a) The total configuration

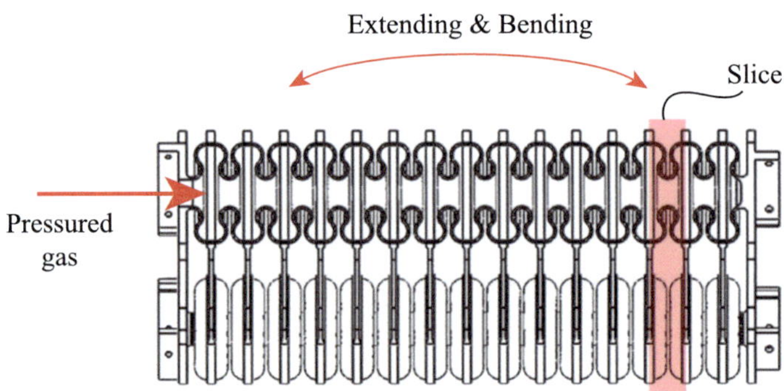

(b) The inflatable control principle

Fig. 2.1 Design schematic of the pneumatic soft actuator

2.1.2 Kinematic Model of the Soft Actuator

It is assumed that there are N slices in the pneumatic soft actuator, shown in Fig. 2.1b. The slice is the basic unit of the soft actuator. The length of the slice is related to the pressure of the input gas. The mathematical model of the slice is shown in Fig. 2.2. The horizontal distance of the cross-section center of two inflatable bellows is marked a, the length of the slice is L_0, the elastic stiffness of the bellows is k, the input pressures are separately P_1, P_2, P_3. The stretch lengths of three slices are ΔL_1, ΔL_2, ΔL_3. Thus:

$$\begin{aligned} P_1 S &= k\Delta L_1 \\ P_2 S &= k\Delta L_2 \\ P_3 S &= k\Delta L_3 \end{aligned} \tag{2.1}$$

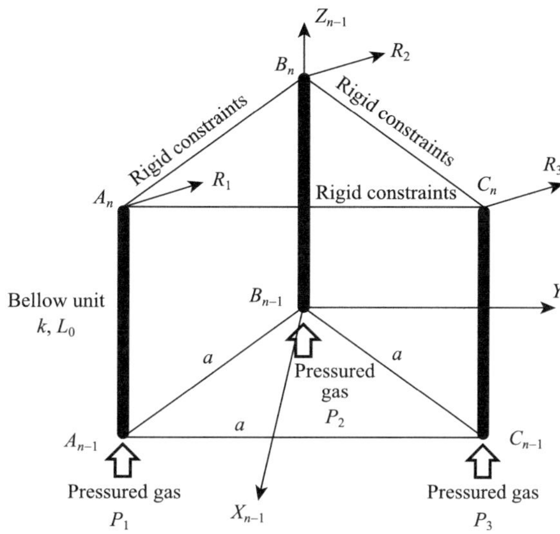

Fig. 2.2 The mathematical model of the pneumatic soft actuator

where, S is the cross-sectional area of the bellow. Assuming that there are no rigid constraints between the bellows, their lengths are described separately as:

$$\begin{aligned} L_1 &= L_0 + \Delta L_1 = L_0 + P_1 S/k \\ L_2 &= L_0 + \Delta L_2 = L_0 + P_2 S/k \\ L_3 &= L_0 + \Delta L_3 = L_0 + P_3 S/k \end{aligned} \tag{2.2}$$

The rigid plate constraints have the following two mechanical effects: keeping three cross-sectional centers of the bellows forming as an equilateral triangle and keeping three cross-sections having the same normal direction.

In Fig. 2.2, $A_{n-1}, B_{n-1}, C_{n-1}$ are the cross-sectional center of three bellows. Using B_{n-1} as the origin point, the body-fixed coordinate $X_{n-1}Y_{n-1}Z_{n-1}$ is defined, shown in Fig. 2.2. $B_{n-1}Y_{n-1}$ is parallel to $A_{n-1}C_{n-1}$. The acting force of the rigid plates to the bellows are in the cross section and described as R_1, R_2, R_3. The generated deflections are described as:

$$v_1 = \frac{R_1 L_1^3}{3E_1 I} = \frac{R_1 L_1^2 S'}{3kI}, \quad v_2 = \frac{R_2 L_2^3}{3E_2 I} = \frac{R_2 L_2^2 S'}{3kI}, \quad v_3 = \frac{R_3 L_3^3}{3E_3 I} = \frac{R_3 L_3^2 S'}{3kI} \tag{2.3}$$

where, E is the elastic modulus; I is the inertia moment and EI is the bending stiffness; S' is the cross-sectional area of the bellow's shell, which is the effective area to afford the tension load. The rotation angles generated by the forces R_1, R_2, R_3 are described as:

$$
\begin{aligned}
\theta_1 &= \frac{R_1 L_1^2}{2E_1 I} = \frac{R_1 L_1 S'}{2kI} \\
\theta_2 &= \frac{R_2 L_2^2}{2E_2 I} = \frac{R_2 L_2 S'}{2kI} \\
\theta_3 &= \frac{R_3 L_3^2}{2E_3 I} = \frac{R_3 L_3 S'}{2kI}
\end{aligned} \tag{2.4}
$$

The cross sections of three bellows have the same normal direction, it means that the projections of three bellows in the X_{n-1}-Y_{n-1} plane are parallel. Thus:

$$
\{\theta_1 = \theta_2;\ \ \theta_1 = \theta_3\} \Rightarrow \{R_1 L_1 = R_2 L_2;\quad R_1 L_1 = R_3 L_3\} \tag{2.5}
$$

The projection angle of the bellows, according to $B_{n-1}Y_{n-1}$ axis, is assumed as α. When the bellows is inflated, the coordinates of A'_n, B'_n, C'_n are given by:

$$
\begin{cases}
A'_n : \left(v_1 \cos\alpha + \dfrac{\sqrt{3}}{2} a \ \ v_1 \sin\alpha - \dfrac{a}{2} \ \ L_1 \right) \\
B'_n : \left(v_2 \cos\alpha \ \ v_2 \sin\alpha \ \ L_2 \right) \\
C'_n : \left(v_3 \cos\alpha + \dfrac{\sqrt{3}}{2} a \ \ v_3 \sin\alpha + \dfrac{a}{2} \ \ L_3 \right)
\end{cases} \tag{2.6}
$$

According to the constraints $A'_n B'_n = B'_n C'_n = C'_n A'_n = a$, there is

$$
\left(\frac{S'}{3kI}\right)^2 \left(R_1 L_1^2 - R_2 L_2^2\right)^2 + \frac{\left(R_1 L_1^2 - R_2 L_2^2\right) a S'}{3kI} \left(\sqrt{3}\cos\alpha - \sin\alpha\right) + (L_1 - L_2)^2 = 0 \tag{2.7}
$$

$$
\left(\frac{S'}{3kI}\right)^2 \left(R_3 L_3^2 - R_2 L_2^2\right)^2 + \frac{a S'}{3kI} \left(R_3 L_3^2 - R_2 L_2^2\right) \left(\sqrt{3}\cos\alpha + \sin\alpha\right) + (L_3 - L_2)^2 = 0 \tag{2.8}
$$

$$
\left(\frac{S'}{3kI}\right)^2 \left(R_3 L_3^2 - R_1 L_1^2\right)^2 + \frac{2a S'}{3kI} \left(R_3 L_3^2 - R_1 L_1^2\right) \sin\alpha + (L_3 - L_1)^2 = 0 \tag{2.9}
$$

It is defined that:

$$
\begin{aligned}
W &= R_1 L_1 = R_2 L_2 = R_3 L_3 \geq 0 \\
M &= S'/(3kI) > 0
\end{aligned} \tag{2.10}
$$

Combining the Eq. (2.5), Eqs. (2.7)–(2.9):

(1) when $\Delta L_1 \neq \Delta L_3$, there is:

$$W = \frac{a \sin \alpha}{M(\Delta L_1 - \Delta L_3)} \pm \frac{1}{M} \operatorname{sign}(\Delta L_1 - \Delta L_3) \cdot \sqrt{\frac{3a^2 - 3(\Delta L_1 - \Delta L_3)^2 - (2\Delta L_2 - \Delta L_3 - \Delta L_1)^2}{3(\Delta L_1 - \Delta L_3)^2 + (2\Delta L_2 - \Delta L_3 - \Delta L_1)^2}} \tag{2.11}$$

(2) when $\Delta L_1 = \Delta L_3 \neq \Delta L_2$, there is:

$$W = \frac{-\sqrt{3}a \cos \alpha}{M(\Delta L_1 - 2\Delta L_2 + \Delta L_3)} \pm \frac{1}{M} \operatorname{sign}(\Delta L_1 - 2\Delta L_2 + \Delta L_3) \cdot \sqrt{\frac{3a^2 - 3(\Delta L_1 - \Delta L_3)^2 - (2\Delta L_2 - \Delta L_3 - \Delta L_1)^2}{3(\Delta L_1 - \Delta L_3)^2 + (2\Delta L_2 - \Delta L_3 - \Delta L_1)^2}} \tag{2.12}$$

where the projection angle α satisfies:

$$\tan \alpha = \frac{\sqrt{3}(L_1 - L_3)}{2L_2 - L_3 - L_1} \tag{2.13}$$

When the input pressures of three bellows P_1, P_2, P_3 are known, the stretch lengths L_1, L_2, L_3 can be obtained. Then R_1, R_2, R_3, α can be calculated from Eqs. (2.11)–(2.13), and the deflections v_1, v_2, v_3 can be calculated by Eq. (2.3). Finally, the coordinates of A'_n, B'_n, C'_n can be calculated by Eq. (2.6). The orientation of the slice actuated by the pressured gas is obtained.

The pneumatic soft actuator is made up of a series of slices stacked along its central axis. According to the constant curvature assumption[3, 4], the slice n has the same curvature as the slice $n - 1$ ($n \in [2, N]$). It is assumed that the angle between the normal directions of the adjacent cross-sections is θ. Then the transfer between two adjacent body-fixed coordinates is obtained and shown in Fig. 2.3.

The transfer matrix between the two body-fitted coordinates is calculated by:

$$\begin{pmatrix} x_{n-1} \\ y_{n-1} \\ z_{n-1} \end{pmatrix} = \boldsymbol{H}^{\mathrm{T}} \begin{pmatrix} x_n \\ y_n \\ z_n \end{pmatrix} + \overrightarrow{B_{n-1}B_n} \tag{2.14}$$
$$\boldsymbol{H} = \boldsymbol{M}_3(\alpha)\,\boldsymbol{M}_2(\theta)\,\boldsymbol{M}_3(-\alpha)$$

where, $\overrightarrow{B_{n-1}B_n}$ is the vector in the coordinate $X_{n-1}Y_{n-1}Z_{n-1}$. It is assumed that the vector of the cross-section center of the first slice in the first body-fixed coordinate $X_0Y_0Z_0$ is assumed as r_1. $\overrightarrow{B_0B_1}$ is the transfer vector between the beginning two

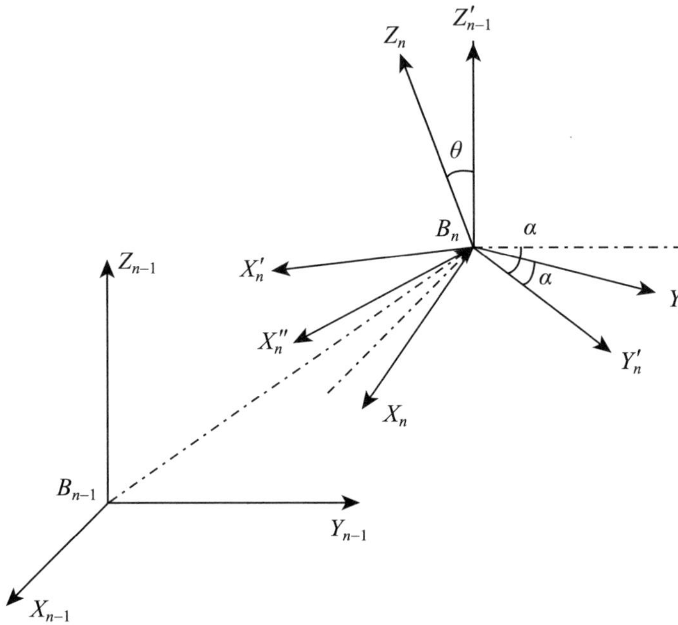

Fig. 2.3 The transfer between two adjacent body-fixed coordinates

body-fixed coordinates. There are totally N slices in the soft actuator. Then, the vector of the cross-section center of the soft actuator r_n can be calculated by:

$$\boldsymbol{r}_n = \left(\boldsymbol{H}^{\mathrm{T}}\right)^{n-1} \boldsymbol{r}_1 + \left[\left(\boldsymbol{H}^{\mathrm{T}}\right)^{n-2} + \left(\boldsymbol{H}^{\mathrm{T}}\right)^{n-3} + \cdots + \left(\boldsymbol{H}^{\mathrm{T}}\right)^{1} + \left(\boldsymbol{H}^{\mathrm{T}}\right)^{0}\right] \overrightarrow{B_0 B_1} \tag{2.15}$$

$$\overrightarrow{B_0 B_1} = \left(v_2 \cos\alpha \;\; v_2 \sin\alpha \;\; L_2 \right)^{\mathrm{T}} \tag{2.16}$$

where, $2 \leq n \leq N$.

2.1.3 Determination of Elastic Stiffness of the Soft Actuator

The prototype of the pneumatic soft actuator is manufactured using 3D print technology [5]. The material is nylon reinforced by glass fiber. This kind of powder mixture is printed using the selective laser sintering (SLS) technology [6]. The elastic stiffness of the soft actuator k, which is necessary to the kinematic model [7, 8], is unknown. In order to get the elastic stiffness k, the proportional relation between the gas pressure and the elongation of the soft actuator needs to be determined. The inputting gas pressure is known and controlled by the proportional valve. The elongation of the soft actuator is measured by the laser rangefinder. Thus, the experiment

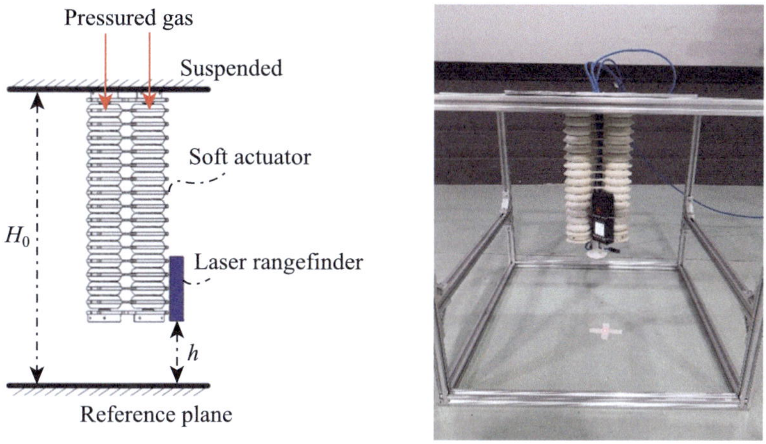

(a) Schematic frame (b) Physical system

Fig. 2.4 Experimental set up used to measure the elastic stiffness

set up used to measure the elastic stiffness is designed and developed, shown in Fig. 2.4.

In Fig. 2.4a, the schematic frame of the experiment is presented. It can be seen that the pneumatic soft actuator is suspended. The height of the frame is assumed H_0. The distance from the free side of the actuator to the reference ground is h. The pressured gas is inputted from the fixed side of the soft actuator. The laser rangefinder is mounted along the surface of the actuator. In Fig. 2.4b, the physical system is shown. The same gas is controlled and allowed to flow into three inflatable bellows. The elongation of the bellows can be measured using the laser rangefinder. It is assumed that the height of the free side at the initial stage (gas pressure relative to surrounding air pressure is zero) is h_0. The height becomes h under the pressure P, then the length increment of the soft actuator is $h_0 - h$. Assuming that the external diameter of the bellow is d, therefore:

$$3P\pi\left(\frac{d}{2}\right)^2 = k\left(h_0 - h\right) \tag{2.17}$$

leading to,

$$h_0 - h = \frac{3\pi}{k}\left(\frac{d}{2}\right)^2 P = CP \tag{2.18}$$

with C is the fitting coefficient, then the elastic stiffness of the bellows k can be calculated by:

$$k = \frac{3\pi}{C}\left(\frac{d}{2}\right)^2 \tag{2.19}$$

Table 2.1 Distance of the free side relative to the reference plane (1bar = 10^5 Pa)

Pressure/bar	0.00	0.20	0.25	0.30	0.35	0.40
h/mm	212	205	204	203	201	200
Pressure/bar	0.45	0.50	0.55	0.60	0.65	0.70
h/mm	198	197	195	194	192	190
Pressure/bar	0.75	0.80	0.85	0.90	0.95	1.00
h/mm	189	187	186	185	182	180

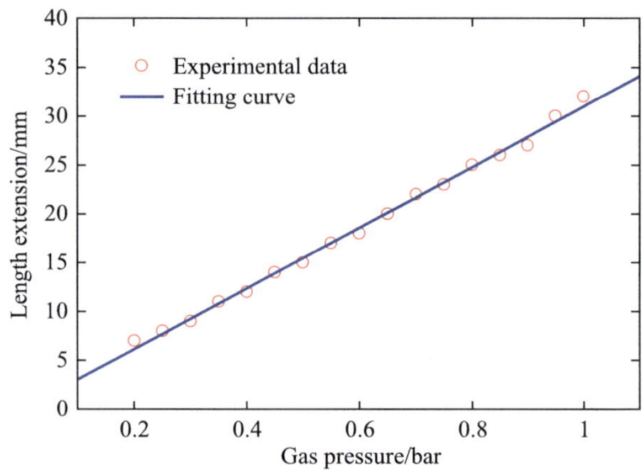

Fig. 2.5 Fitting curve of the actuation length versus input pressure

The external diameter of the bellow is known as $d = 50$mm and the free length of the soft actuator is 306mm. When the soft actuator is inflated with different pressure, the distance of the free side of the actuator relative to the reference plane is measured and is listed in Table 2.1.

According to the data of Table 2.1, the best-fit curve is obtained, as shown in Fig. 2.5. It can be seen that the fitting coefficient $C = 3.1017 \times 10^{-7}$m/Pa, giving the elastic stiffness of the bellows $k = 1.8991 \times 10^4$ N/m.

2.2 Design and Development of the Soft Manipulator

2.2.1 Kinematic Model of the Soft Manipulator

The pneumatic soft manipulator is assembled with three soft actuators (section-1 to section-3), shown in Fig. 2.6. Each actuator has three inflatable bellows, thus there are totally nine bellows in the soft manipulator. The numbering rule of the bellows can be found in Fig. 2.6.

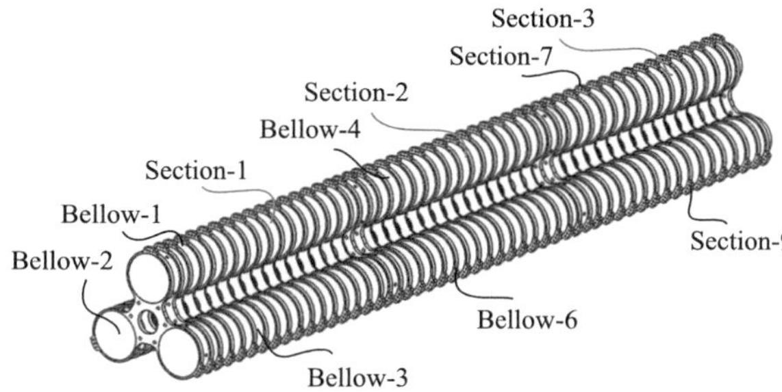

Fig. 2.6 The assemble of the soft manipulator

In the body-fixed coordinate of each soft actuator, the vector of the cross-section center can be calculated by Eqs. (2.15) and (2.16). It is assumed that the vector of the second actuator in its body-fixed coordinate is $(r_n)_2$. It should be transferred into the body-fixed coordinate of the first section. The vector transformation is shown in Eq. (2.20):

$$(r_n)_{2-1} = \left(\boldsymbol{H}_1^{\mathrm{T}}\right)^N \left[(r_n)_2 + \begin{pmatrix} 0 \\ 0 \\ T_{1\text{-}2} \end{pmatrix} \right] + \left[\left(\boldsymbol{H}_1^{\mathrm{T}}\right)^{N-1} + \left(\boldsymbol{H}_1^{\mathrm{T}}\right)^{N-2} + \cdots + \left(\boldsymbol{H}_1^{\mathrm{T}}\right)^1 + \left(\boldsymbol{H}_1^{\mathrm{T}}\right)^0 \right] \left(\overrightarrow{B_0 B_1}\right)_1 \quad (2.20)$$

where, $\boldsymbol{H}_1 = \boldsymbol{H}\,(\alpha_1, \theta_1), \alpha_1$ and θ_1 are the relative angle in the first actuator; $\left(\overrightarrow{B_0 B_1}\right)_1$ is also the vector calculated in the body-fixed coordinate of the first section. $T_{1\text{-}2}$ is the thickness of the connecting part between two actuators. Similarly, the vector transformation is shown in Eq. (2.21):

$$(r_n)_{3\text{-}2} = \left(\boldsymbol{H}_1^{\mathrm{T}}\right)^N \left[(r_n)_3 + \begin{pmatrix} 0 \\ 0 \\ T_{2\text{-}3} \end{pmatrix} \right] + \left[\left(\boldsymbol{H}_2^{\mathrm{T}}\right)^{N-1} + \left(\boldsymbol{H}_2^{\mathrm{T}}\right)^{N-2} + \cdots + \left(\boldsymbol{H}_2^{\mathrm{T}}\right)^1 + \left(\boldsymbol{H}_2^{\mathrm{T}}\right)^0 \right] \left(\overrightarrow{B_0 B_1}\right)_2 \quad (2.21)$$

where, $\boldsymbol{H}_2 = \boldsymbol{H}\,(\alpha_2, \theta_2)$, α_2 and θ_2 are the relative angle in the second actuator; $\left(\overrightarrow{B_0 B_1}\right)_2$ is also the vector calculated in the body-fixed coordinate of the second section. $T_{2\text{-}3}$ is the thickness of the connecting part between sections 2 and 3. Then the vector transformation from section-3 to section-1 can also be derived. Finally, the kinematic model [9] of the soft manipulator is developed.

Table 2.2 The parameters of the soft manipulator

Parameter	L_0	a	N
Value	0.02m	0.075m	16
Parameter	d_{out}	d_{in}	T
Value	0.05m	0.0488m	0.02m

2.2.2 *Simulation and Analysis*

In the kinematic model, the key parameters of the soft manipulator are listed as follows: the length of the slice L_0, the horizontal distance of the cross-section center of two inflatable bellows a, the external diameter of the bellow d_{out} and the inner diameter of the bellow d_{in}, the number of the slice in each bellow N, the thickness of the connection part between two actuators T. These parameters are listed in Table 2.2.

2.2.2.1 Numerical Example I

In this example, the bellows 1, 4 and 7 are inflated with the same gas pressure. The pressure is 0–1 bar, with 0.2 bar being the interval. The relative pressure of the other bellows is zero (relative to the environment pressure). The shape of the soft manipulator changes with the increase of the pressure, as shown in Fig. 2.7.

In Fig. 2.7, the shape deformation of the soft manipulator in five pressure cases is shown. It is seen that the shapes of the soft manipulator in these five cases are consistently curving, which can be found in Fig. 2.7a. It is because the bellows 1, 4 and 7 are along the same central axis. As the flow of the pressured gas is continued, the lengths of the three bellows continue to extend. The pressure of the other six bellows relative to the surrounding environment keeps zero. Under the combination of the rigid constraints, the prolongation of the three bellows leads to the compression of the other six bellows. The overall movement of the soft manipulator is elongated and curving. The bending amplitude increases with the increase of the gas pressure. In Fig. 2.7b, the curves in Fig. 2.7a are projected in the X-Y plane. It is seen that the projections of the curves in the X-Y plane is a straight line. It is because the three bellows of the soft actuator are distributed in three points of an equilateral triangle. One of the three bellows extends, its projection in the X-Y plane is perpendicular to the opposite line of the equilateral triangle. Three bellows along the same central axis extend, the projections in the X-Y plane should be in the same straight line. This feature found in Fig. 2.7b validates the reasonability of the kinetic model of the soft manipulator.

Fig. 2.7 Shape deformation of the soft manipulator in the first example

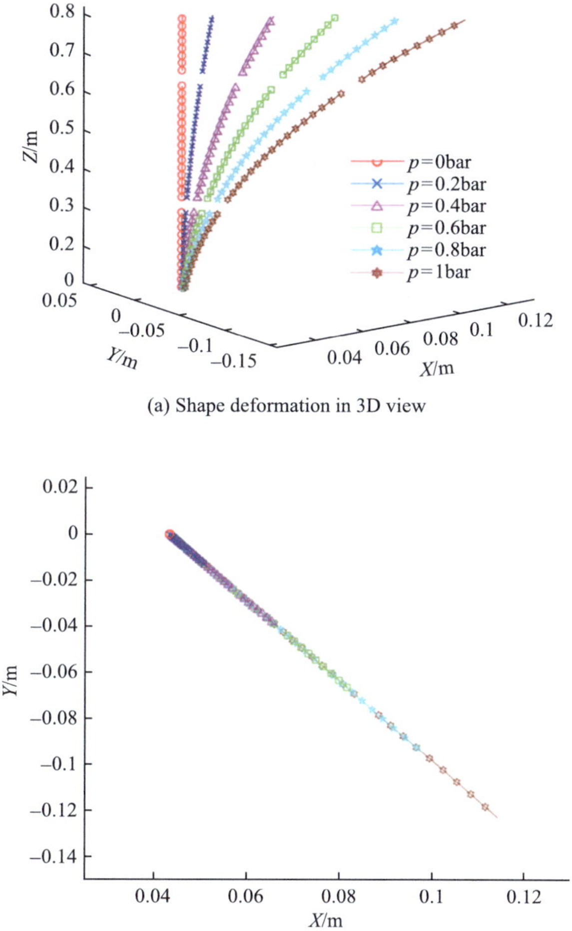

(a) Shape deformation in 3D view

(b) Projection in the X-Y plane

2.2.2.2 Numerical Example II

In the second example, the bellows 1, 5, 6 and 7 are inflated with the same pressured gas. The pressure is 0–1 bar, with 0.2 bar being the interval. The pressure of the other bellows relative to the surrounding environment is zero. The shape of the soft manipulator changes as the increase of the pressure, shown in Fig. 2.8.

In Fig. 2.8a, the shape deformation of the soft manipulator in five pressure cases is shown. It is seen that the shapes of the soft manipulator in these five cases are close to the "S" style. It is because the bellows 5 and 6 are not along the central axis of the bellows 1 and 7. The three bellows 4, 5 and 6 are in the same soft actuator (section-2 in Fig. 2.6), the bending effect of the bellows 5 and 6 with the same gas

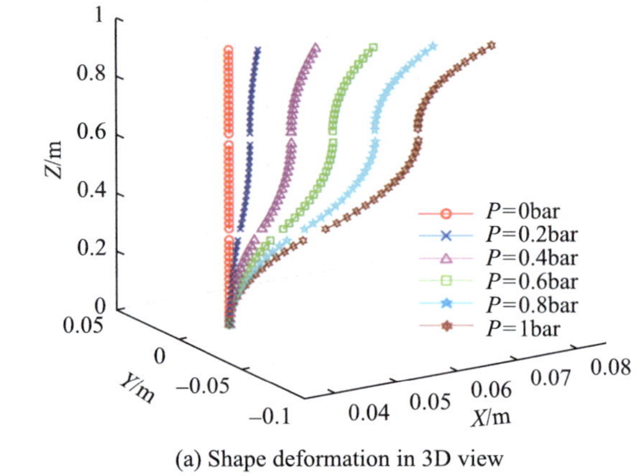

Fig. 2.8 Shape deformation of the soft manipulator in the second example

pressure is just opposite to the bellow-4. When the four bellows (1, 5, 6 and 7) are inflated with the same pressure, the middle section-2 bends to the opposite direction relative to the other two sections (section-1 and section-3 in Fig. 2.6). The overview of the soft manipulator shows "S" shape. The deformation amplitude increases with the increase of the gas pressure. In Fig. 2.8b, the curves in Fig. 2.8a are projected in X-Y plane. It is also seen that the projections of the curves in X-Y plane is a straight line. It is because the projections of the bellows 5 and 6 in X-Y plane is parallel to the projections of the bellows 1 and 7. Thus, the projections of the central lines of three soft actuators in X-Y plane should be in the same straight line. This feature found in Fig. 2.8b validates the reasonability of the kinetic model of the soft manipulator again.

2.3 Experiment Analysis

2.3.1 Development of the Experiment System

The experiment system includes the soft manipulator, the pneumatic control system and the position measurement system. The soft manipulator is composed of three soft actuators, which are manufactured by SLS print technology. In addition, a soft gripper is designed and mounted on the end of the soft manipulator. There are totally nine bellows in the soft manipulator, thus, nine independent pneumatic pipelines are necessary. The soft gripper is also controlled by the pressured gas. The open and close actions need two independent pneumatic pipelines. Therefore, the pneumatic control system needs eleven independent pipelines. Its system schematic and design are shown in Fig. 2.9.

The pneumatic control system is mounted in the satellite simulator. The pressure in the pipelines is controlled using FESTO proportional valve. The pressure injected into the nine bellows is limited to 0–2 bar and the pressure in the soft gripper is limited to 0~5 bar. Therefore, the bellows and the gripper are inflated using two input pressure, shown in Fig. 2.9. The inputting pressure of the bellows is set to 2 bar and the inputting pressure of the soft gripper is set to 5 bar. Accordingly, there are two reducing valves and two pressure sensors in the pneumatic control system. The arrangement of this pneumatic hardware is shown in Fig. 2.10a. The proportional valves and the pressure sensors are mounted on the internal surfaces of the side plates of the satellite simulator. The blue pipes are the pressured gas

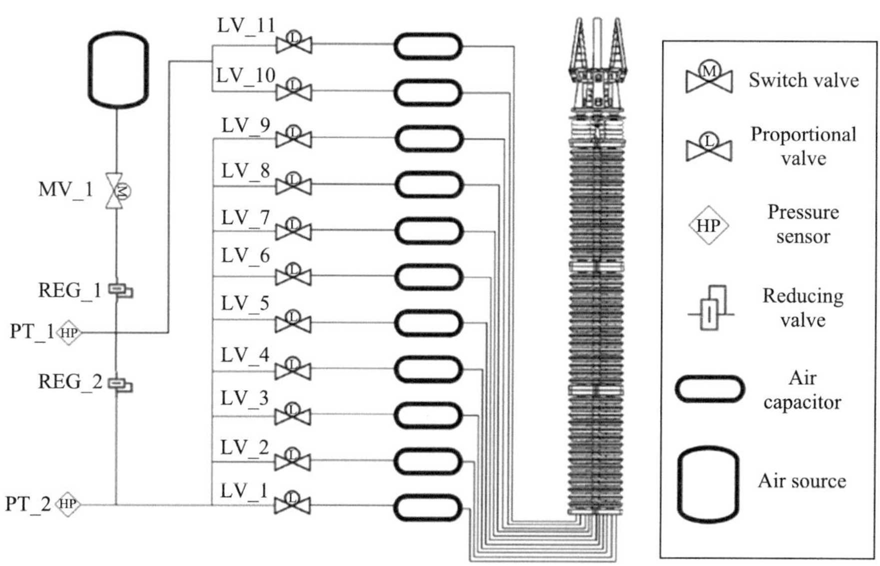

Fig. 2.9 The design of the pneumatic control system

Fig. 2.10 The experiment system of the soft manipulator

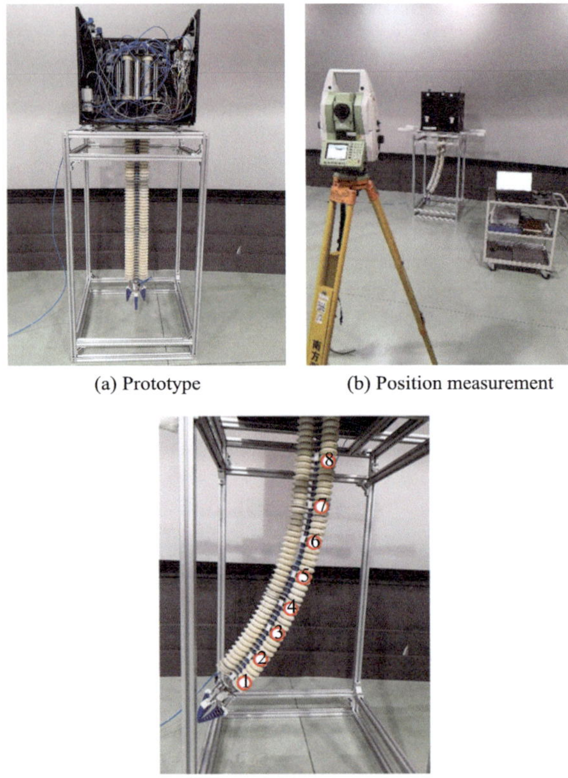

(a) Prototype (b) Position measurement

(c) Markers settlement

pipelines, and the gray cables are the wires of the valves and sensors. The design schematic shown in Fig. 2.9 is the guide for the connection of the gas pipelines and the wires.

In order to obtain the shape deformation and end position of the soft manipulator, the position measurement system is designed and developed using Leica total station TS30, shown in Fig. 2.10b. Eight markers are inscribed along the central axis of the soft manipulator, shown in Fig. 2.10c. The first marker is inscribed around the end position of the soft manipulator. The displacements of these markers in the coordinate of the total station can be measured using TS30. Then the shape deformation and end position of the soft manipulator can be obtained.

Combining the soft manipulator, the pneumatic control system and the position measurement system, the experiment system of the soft manipulator is developed, shown in Fig. 2.10.

In Fig. 2.10a, the prototype includes the soft manipulator and the pneumatic control system. The pneumatic control system is arranged in the satellite simulator, which is mounted on the frame. The soft manipulator is suspended under the satellite simulator. In Fig. 2.10b, it is seen that the total station stands in front of the prototype.

2.3.2 Experimental Results Analysis

The coordinates of the eight markers are measured by the total station TS30. Using these coordinates, the shape deformation of the soft manipulator can be depicted. To validate the correctness of the proposed kinematic model, the pneumatic situation with the bellows 1, 4 and 7 inflated is chosen. In this situation, three cases with different gas pressure are discussed. In the first case, the pressured gas of 0.4 bar is imported to the three bellows 1, 4 and 7. In other two cases, the gas pressure is separately 0.8 and 1.2 bar. The comparison of the shape deformations in these three cases is shown in Fig. 2.11.

Fig. 2.11 Shape deformations obtained by the model and the experiment

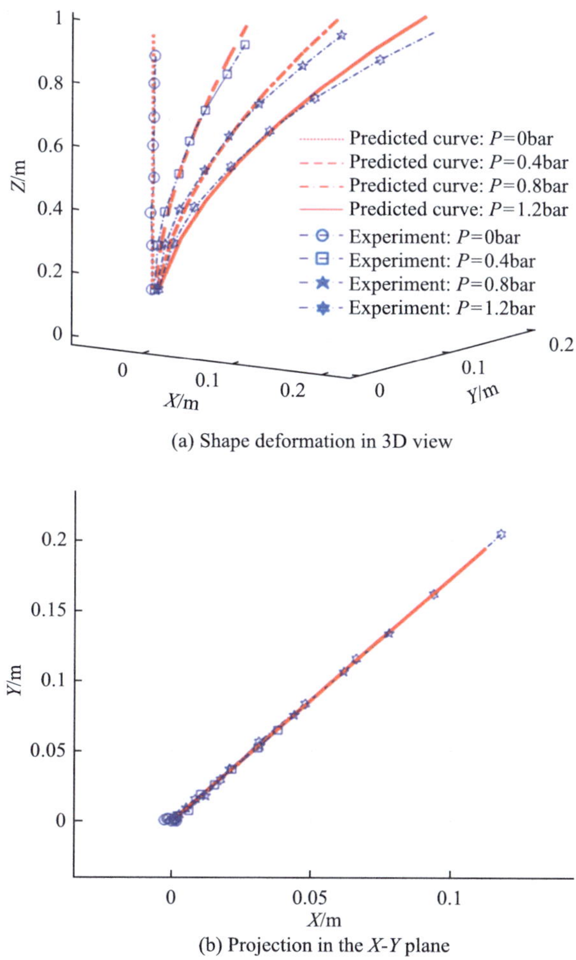

Figure 2.11a shows a comparison of the shape deformation in three cases in the 3D view. It can be seen that the predicted curves show good agreement with the depicted curves obtained by the experiment. Figure 2.11b shows that the projections of all these curves in the X-Y plane are in the same straight line, which verifies the motion stability of the soft manipulator.

To evaluate the end-repeat positioning accuracy [10] of the soft manipulator, three actuation cases with the same gas pressure (1.2 bar) are selected. In the first case, the bellows 1, 4 and 7 are inflated with pressured gas 1.2 bar. In the second case, the bellows 2, 3, 5, 6, 8 and 9 are inflated with pressured gas 1.2 bar. In the third case,

Fig. 2.12 The end-repeat positioning accuracy in three cases

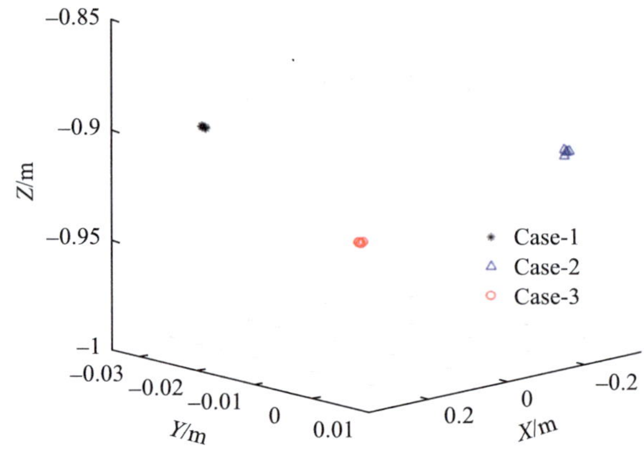

(a) The end positions of the soft manipulator in three cases

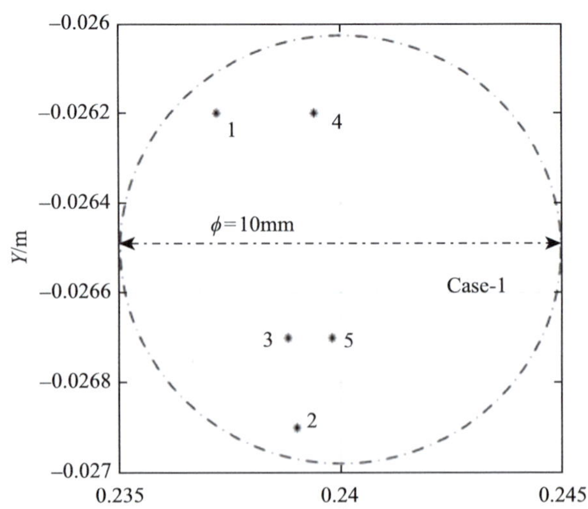

(b) X-Y plane projection of five points in Case-1

Fig. 2.12 (continued)

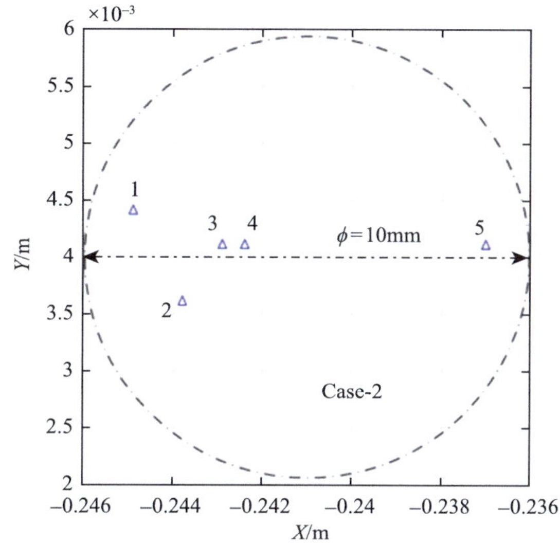

(c) *X-Y* plane projection of five points in Case-2

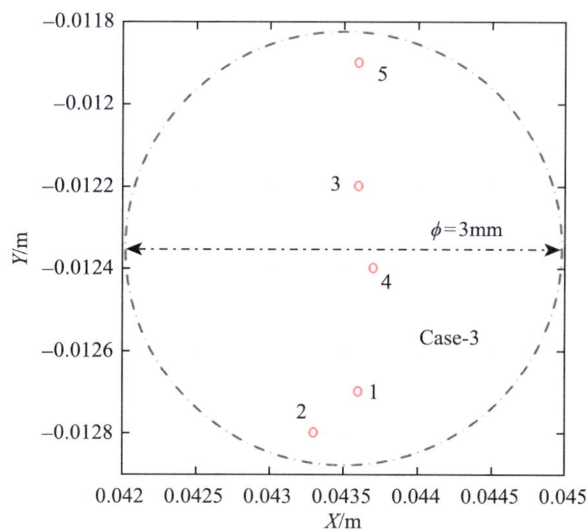

(d) *X-Y* plane projection of five points in Case-3

the bellows 1, 5, 6 and 7 are inflated with pressured gas 1.2 bar. In each case, the bellows are inflated and the end position of the soft manipulator (the coordinate of the first marker) is measured. Then the bellows are deflated. This inflating-measuring-deflating process is repeated five times. The experimental results were recorded and shown in Fig. 2.12.

Figure 2.12a shows the end positions of the soft manipulator in three cases in 3D view. It can be seen that the five points in each case almost overlap. It means that the end position of the soft manipulator keeps a specific and stable relation with the gas pressure and inflating pattern. In other words, the soft manipulator has a stable end-repeat positioning accuracy. In order to show these points clearly, the three cases in Fig. 2.12a are enlarged and shown separately in Fig. 2.12b–d. The five points in Case-1, which are numbered according to their recording order in the experiment, are projected in the X-Y plane and shown in Fig. 2.12b. It indicates that these five points can be included in a circle with a diameter 10mm. The five points in Case-2 are projected in the X-Y plane and shown in Fig. 2.12c. These five points can also be included in a 10mm circle. The five points in Case-3 are projected in the X-Y plane and shown in Fig. 2.12d. It indicates that these five points can be included in a 3mm circle. From above experiment results and analysis, it is concluded that the end-repeat positioning accuracy of the soft manipulator is better than 10mm.

2.4 Summary

In order to develop a soft manipulator, a practical design must consider actuation mode, structure design, material, manufacture, theoretical modeling and control. In this chapter, a novel design of a pneumatic soft manipulator is proposed. The soft manipulator is actuated using pressured gas. Different actuation bellows and gas pressure lead to different movement of the soft manipulator. Due to the complex structure of the soft manipulator, it is too difficult to manufacture using the traditional machining method. Therefore, the 3D printing technology using composite material is adopted. From the experimental results, it is known that the soft manipulator is able to move properly under the actuation of the pressured gas. The structure of the soft actuator printed using nylon reinforced with glass fiber has good tenacity and stability. To develop the kinematic model of the soft manipulator, which is the basis of its motion control, the piecewise constant curvature method is introduced. The comparison of the shape deformation obtained from the theoretical model and the experiment indicates that the proposed model can give a good motion prediction of the soft manipulator. Next, the motion control with the position feedback will be developed. The structure optimization of the soft manipulator will focus on the length, the thickness and the distribution of the bellows and the thin-rigid constraint plates.

References

1. Xavier MS, Tawk CD, Zolfagharian A et al (2022) Soft pneumatic actuators: a review of design, fabrication, modeling, sensing, control and applications. IEEE Access 10:59442–59485
2. Bishop-Moser J, Kota S (2015) Design and modeling of generalized fiber-reinforced pneumatic soft actuators. IEEE Trans Robot 31(3):536–545

3. Falkenhahn V, Mahl T, Hildebrandt A et al (2014) Dynamic modeling of constant curvature continuum robots using the euler-lagrange formalism. IEEE/RSJ international conference on intelligent robots and systems. IL, USA, Chicago, pp 2428–2433
4. Della Santina C, Bicchi A, Rus D (2020) On an improved state parametrization for soft robots with piecewise constant curvature and its use in model based control. IEEE Robot Autom Lett 5(2):1001–1008
5. Bozkurt Y, Karayel E (2021) 3D printing technology; methods, biomedical applications, future opportunities and trends. J Market Res 14:1430–1450
6. Charoo NA, Barakh Ali SF, Mohamed EM et al (2020) Selective laser sintering 3D printing–an overview of the technology and pharmaceutical applications. Drug Dev Ind Pharm 46(6):869–877
7. Armanini C, Boyer F, Mathew AT et al (2023) Soft robots modeling: a structured overview. IEEE Trans Rob 39(3):1728–1748
8. Wang J, Chortos A (2022) Control strategies for soft robot systems. Adv Intell Syst 4(5):2100165
9. Bamdad M, Bahri MM (2019) Kinematics and manipulability analysis of a highly articulated soft robotic manipulator. Robotica 37(5):868–882
10. Chen X, Zhang X, Liu H et al (2020) Design and development of a soft robotic manipulator. Int J Mech Mater Des 16(2):309–321

Chapter 3
Design and Optimization of a Pneumatic Soft Actuator

The performance of a pneumatic manipulator is closely related to each design parameter. Establishing a parametric model of the manipulator and selecting high-sensitivity parameters through the sensitivity analysis of all the parameters are essential for improving the overall performance of the pneumatic manipulator. The corresponding optimization algorithm can then be used to optimize the target. Due to the particularity of the structure and material of the pneumatic soft manipulator, the present simulation and optimization method for the traditional airbag is not suitable for the pneumatic soft manipulator. The specific method is needed in this issue to optimize the motion and mechanical performances of this manipulator.

This chapter proposes an optimal design method for a pneumatic soft actuator to solve the problem of the limited inclination angle and the support reaction force. The geometric relationship between structural dimensions is established by analyzing the structural characteristics. A parametric simulation model of a pneumatic soft manipulator is established by Python based on the static balance theory. Moreover, the ABAQUS benchmark model is simulated on this basis, and then the ABAQUS and ISIGHT software are combined to implement the optimization. The bending inclination angle and support reaction force are taken as optimization objectives, the highly sensitive factors are selected, and the Evol optimization algorithm is used to optimize the inclination angle and support reaction force respectively; The inclination angle and support reaction force optimization target are coupled through the normalization and weighting strategy to realize the optimization analysis of inclination and support reaction force coupling target, and the optimal parameter is obtained. The bending deformation capacity and load capacity of the pneumatic soft manipulator are improved.

X. Zhang et al., *Design and Implementation of Soft Robotic Manipulators*,
https://doi.org/10.1007/978-981-95-1657-5_3

Fig. 3.1 Schematic of soft pneumatic manipulator

3.1 Model Parameterization

3.1.1 Geometric Model and Physical Process Description

The pneumatic soft manipulator consists of three identical inflatable bellows symmetrically connected in parallel through a series of thin plates. Figure 3.1 presents the overall design, and Fig. 3.2 presents the sectional drawing and inflation modes. The amplitudes of deformation in length vary when the bellows are inflated at different air pressures. The soft manipulator can realize bending movement due to the combination of inflatable bellows and thin plates.

3.1.2 Structural Parameterization

The structural optimization is divided into two types: parametric and nonparametric [1]. The former uses parametric methods to analyze the response of structural parameters and processing technology parameters to the target quantity, and then obtain the influence law of the factors and the optimal solution. The latter considers the maximum stiffness, minimum mass, etc. as objectives to implement topol-

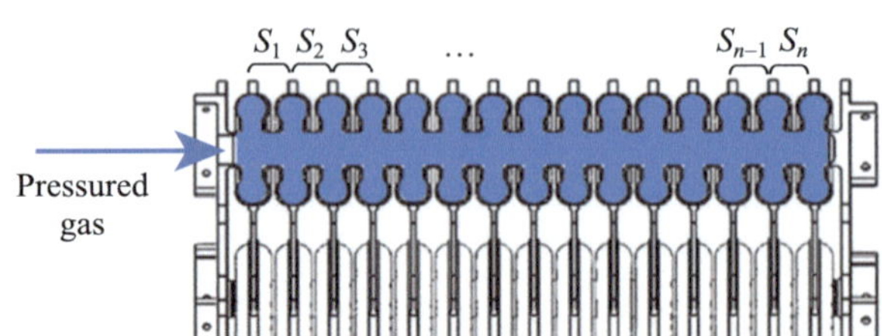

Fig. 3.2 Section view of the soft pneumatic manipulator

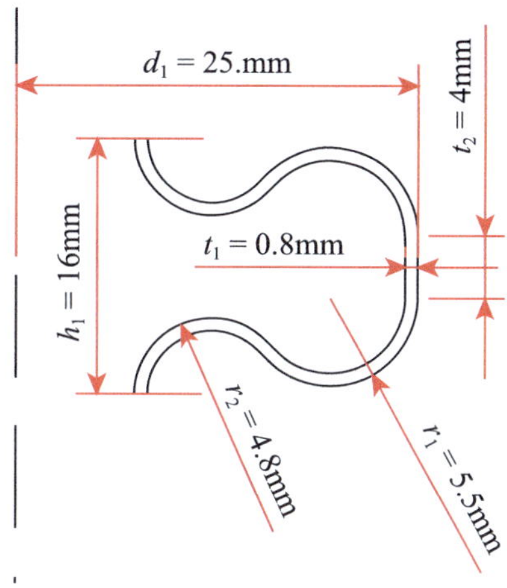

Fig. 3.3 Parameterized diagram of the minimum characteristic structure of the airbag

ogy [2] or morphology optimization [3], which generally changes the shape of the structure significantly. The shape deformation optimization of a pneumatic manipulator is a parametric optimization, where each feature size is controlled during the structural parameterization. The dimensional constraints include vertical, tangential, parallel, equal, and fixed constraints. The structural parameterization process is implemented to avoid structural errors, such as intersection, dislocation, and overlap, and is described in detail in the following section.

The parametrization diagram of the minimum characteristic structure of the airbag is presented in Fig. 3.3, where each dimension is an independent variable for structural optimization.

The parametric dimensions with some of the non-critical parts such as circular holes and fillets removed are presented in Figs. 3.4 and 3.5. During structural optimization, $r_3(= d_1)$, $r_4(= r_3 + 2)$, and $t_4(= t_2)$ are observed to be non-independent variables, whereas the other dimensions are independent variables.

3.2 Structural Analysis and Optimization Design Method

3.2.1 Structural Analysis Method

The nonlinear finite element algorithm is used to simulate and analyze the deformation process of the pneumatic soft manipulator, and the programmed automatic pre-processing and post-processing operations are realized through Python programming. This process is comprehensively described in the following subsections.

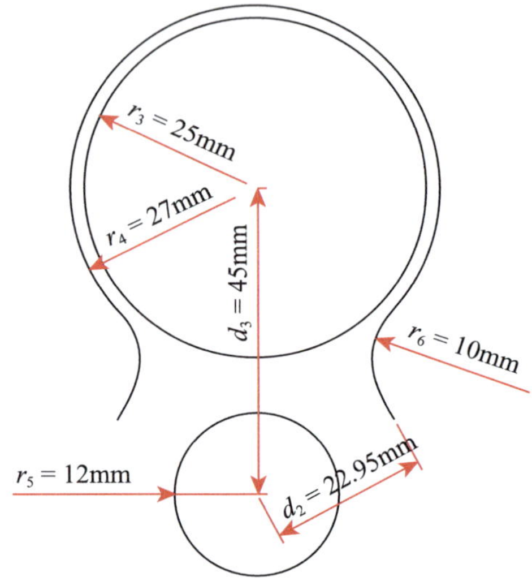

Fig. 3.4 Parameterized plan of the smallest characteristic structure of the thin plate

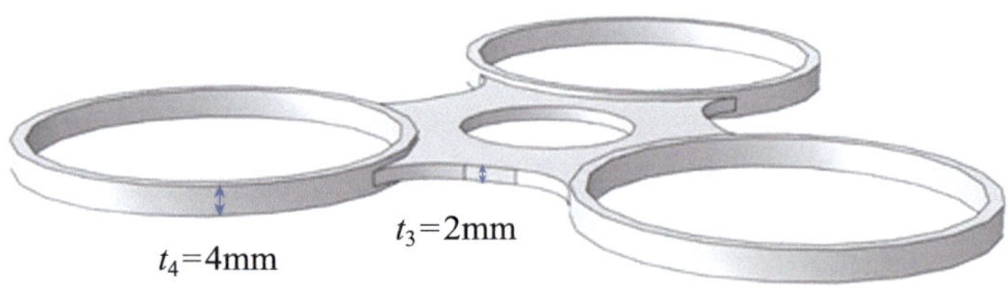

Fig. 3.5 Parametric stereogram of the minimum characteristic structure of the thin plate

3.2.1.1 Geometric Modeling

The finite element simulation modeling method [4, 5] is similar to the three-dimensional (3D) CAD software, where the 3D entities are generated through two-dimensional (2D) sketching. The script is developed in Python to establish the parametric model of the pneumatic soft manipulator, as shown in Fig. 3.6. A 3D rotating body airbag is generated through 2D sketching by Python. Figure 3.7a presents the final geometric model of the pneumatic soft manipulator. The optimization and calculation efficiency is improved using half of the model in the actual simulation based on the symmetry of geometry, load, and boundary, as shown in Fig. 3.7b. The dimensional constraint correlation between the structures should be examined carefully to establish parametric modeling. The dimensional constraints include the internal characteristics of the pneumatic soft manipulator structure, which can be considered for structural modeling. Other constraints are imposed to avoid structural errors such as interference, which can be considered as optimization con-

```
s = mdb.models['actuator_sim'].ConstrainedSketch(name='__profile__', sheetSize=500.0)
g, v, d, c = s.geometry, s.vertices, s.dimensions, s.constraints
s.setPrimaryObject(option=STANDALONE)
s.ConstructionLine(point1=(0.0, -250.0), point2=(0.0, 250.0))
s.FixedConstraint(entity=g[2])
s.retrieveSketch(sketch=mdb.models['actuator_sim'].sketches['bellows_sketch'])
session.viewports['Viewport: 1'].view.fitView()
s.sketchOptions.setValues(constructionGeometry=ON)
s.assignCenterline(line=g[2])
p = mdb.models['actuator_sim'].Part(name='bellowsall', dimensionality=THREE_D, type=D
p = mdb.models['actuator_sim'].parts['bellowsall']
p.BaseSolidRevolve(sketch=s, angle=360.0, flipRevolveDirection=OFF)
s.unsetPrimaryObject()
p = mdb.models['actuator_sim'].parts['bellowsall']
session.viewports['Viewport: 1'].setValues(displayedObject=p)
del mdb.models['actuator_sim'].sketches['__profile__']
```

Fig. 3.6 Python parametric modeling example

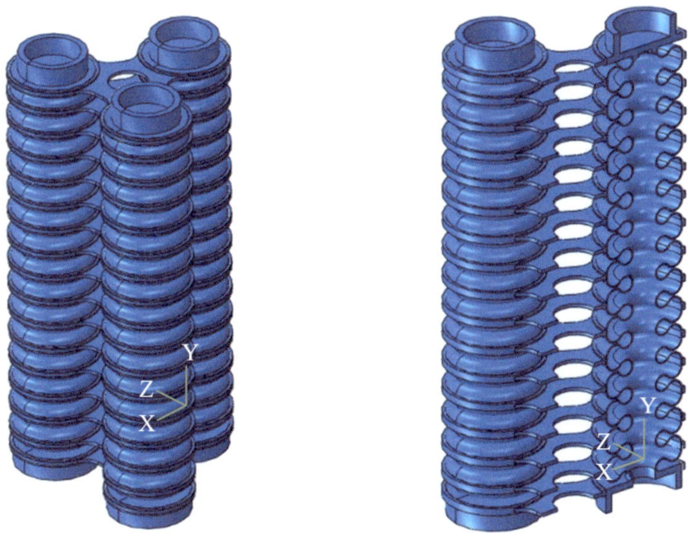

(a) Whole model (b) Semi-model

Fig. 3.7 Simulation geometric model of the soft pneumatic manipulator

straints, as shown in Table 3.1. The automatic control of these two types of constraints should be implemented in the program to facilitate the optimization analysis.

3.2.1.2 Grid Division and Load Setting

The tetrahedral and hexahedral elements should be compared to comprehensively determine the calculation accuracy and the amount of calculation. According to the finite element theory and mechanical theory, the shape function of the first-order tetrahedral element can be considered as a linear function of coordinates.

Table 3.1 Parametric modeling dimensional constraints

Dimensional constraint relationship	
Structural modeling	Optimization constraints
Vertical, parallel, equal, tangential, etc	Avoid interference and intersection, and set the difference or sum of dimensions to be greater than or less than 0

Consequently, the strain obtained by considering the partial derivation is constant, and the stress calculated by the constitutive equation is also constant. Therefore, the first-order tetrahedral element is called a constant stress-strain element. The hexahedral element shape function contains non-primary terms, which present better accuracy. However, a first-order hexahedron element has natural defects. Shear deformation occurs when a first-order complete integral hexahedron element is used [6], owing to the action of an external force, which should not exist because the element edge cannot bend. Furthermore, the element appears to be too rigid. This phenomenon is known as the shear self-locking phenomenon. When a first-order reduced integral hexahedron element is used, the neutral axis fails to bend despite the absence of additional shear deformation, which indicates that the entire element is in the zero-energy mode. Therefore, the element is too soft, and the external force does not produce strain energy; Instead, it produces a large amount of pseudo strain energy. This is known as the hourglass phenomenon. Both of these phenomena may produce incorrect results. The impact of these two phenomena can be reduced by first refining the mesh and increasing the number of meshes because the first-order complete integral element is too hard and the first-order reduced integral element is too soft, as explained earlier. The mesh should be further refined to decrease the stiffness of the first-order complete integration element and increase the stiffness of the first-order reduced integration element. Subsequently, the second-order element can be used to eliminate the influence of the shear self-locking and hourglass phenomena. However, these two methods increase the amount of calculation. The hexahedral element is unsuitable owing to the large amount of computation involved in the structural optimization of the pneumatic soft manipulator. Therefore, a C3D4 tetrahedral element is selected for meshing to ensure that the amount of calculation remains within an acceptable range without compromising the calculation accuracy, as shown in Fig. 3.8.

The structure optimization process of the pneumatic soft manipulator is based on two load conditions:

(1) Load condition one: the bottom of the manipulator is fixed whereas the other end remains free; One inflatable bellow is inflated to realize bending deformation, and the bending angle is noted (as shown in Fig. 3.9).
(2) Load condition two: both ends of the manipulator are fixed; one inflatable bellow is inflated to realize internal extrusion, and the support reaction force at the top of the manipulator is noted.

Fig. 3.8 Mesh generation of simulation geometric model of the soft pneumatic manipulator

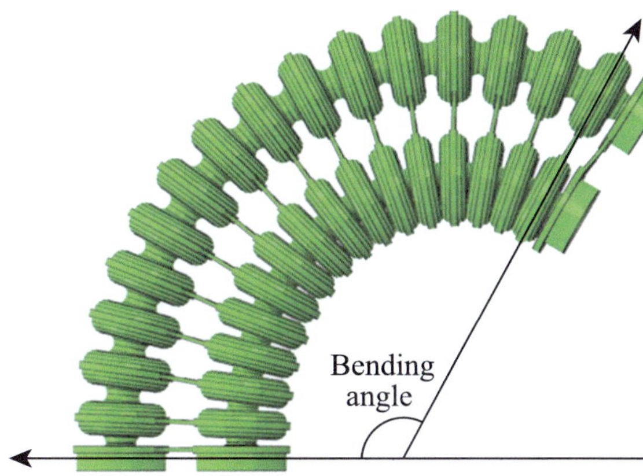

Fig. 3.9 Schematic of bending inclination

The loading process is depicted in Fig. 3.10. The internal pressure finally acts on the inner surface of the airbag in the form of shear, resulting in directional bending.

3.2.1.3 Solver Selection

The implicit calculation performed in the finite element numerical calculation adopts the Newton-Raphson [7] incremental iteration method and uses the partial derivative stress on a strain-tangent stiffness matrix to continuously calculate and update the node displacement and the equivalent node load. The current incremental step calculation is completed when the tolerance convergence condition is satisfied. The explicit algorithm is primarily implemented because the mass matrix used in the

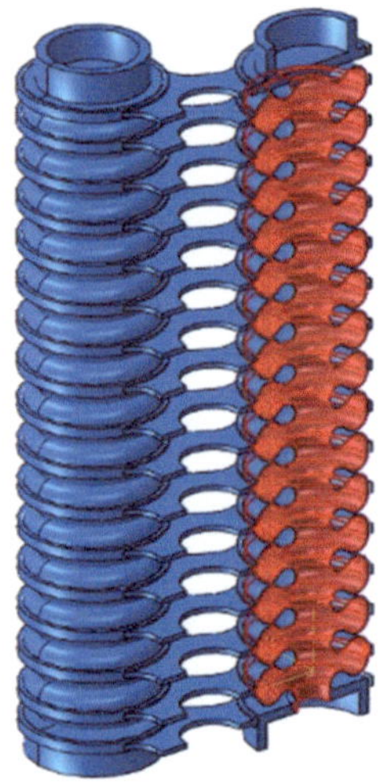

Fig. 3.10 Schematic diagram of pneumatic driver charging and loading

dynamic equilibrium equation is a diagonal concentrated matrix. Therefore, only the diagonal elements are considered in the matrix multiplication, which does not require obtaining the node displacement by solving the inverse matrix of the tangent stiffness matrix, similar to the implicit algorithm. The correlation between the displacement, velocity, and acceleration is a central difference; that is, the displacement and velocity can be expressed by the acceleration equation. Therefore, the acceleration can be obtained firstly from the dynamic balance equation; then, the displacement and velocity can be obtained according to the above relationship. Further, the strain and stress can be obtained from the geometric function matrix and constitutive matrix. The implicit algorithm significantly improves the calculation accuracy using the incremental iterative method. Conversely, the explicit algorithm adopts the central difference method, which is essentially a linear interpolation, and reduces the numerical accuracy, which should be compensated by reducing the time step. The two algorithms cannot be accurately compared in terms of computation. This is because the implicit algorithm adopts an iterative method, and the amount of calculation is based on the degree of convergence; conversely, the explicit algorithm usually requires a large amount of calculation because it involves linearization and requires an extremely small time step to consider the stress wave transmission and possible resonance behavior.

These two algorithms present both advantages and disadvantages. The selection of the methods depends on the physical process in the actual simulation. The problem is generally classified as a static problem when the external force acts for a long time and the speed is infinitely close to zero; a static implicit algorithm is preferred in this case. Conversely, it is classified as a dynamic problem when the action time of the external force is short and the speed is large; an explicit dynamic algorithm is preferred in this case. However, the selection of the algorithm is not absolute, such as low-speed cutting, complex contact, damage, crack propagation, and other processes.

The kinetic energy accounts for a very small proportion, which can be considered as a quasi-static process, but an explicit algorithm should be adopted to solve this problem. This is because an implicit algorithm will cause serious convergence problems and computational difficulties in these processes. Particularly, an implicit dynamic method can also be used for the calculation to obtain more accurate dynamic process results, but the amount of calculation is large.

The analysis and evaluation of the difference between the implicit and explicit algorithms are based on the continuity and calculation amount of the optimization process because the implicit algorithm often produces a convergent result, which interrupts the calculation and is extremely detrimental to the optimization iteration. Conversely, the explicit algorithm is often suitable for high-speed impact processes. If the loading time is extended to achieve a quasi-static process, it will result in a large amount of calculation. The calculation amount of the explicit algorithm is observed to be much larger than that of the implicit algorithm, according to the evaluation of the bending process of the pneumatic manipulator; therefore, the implicit algorithm is selected for optimization.

3.2.2 Optimization Design Method

3.2.2.1 Optimization Method

Optimization algorithms [8] are typically classified into three categories in the Isight software: numerical optimization, direct search optimization, and global exploration optimization algorithms. The numerical optimization algorithm depends on the continuity and derivability of the objective function. Gradient optimization is performed with a certain exploration step based on the given initial value of the independent variable. The calculation process is efficient and fast, but it easily falls into a local optimum owing to an iterative algorithm. Conversely, the direct search optimization algorithm does not require the continuity or derivability of the objective function or even the description of the function expression. The exploration step is large, and the calculation process is more efficient. However, it also easily falls into the local optimum because an iterative algorithm is used in this algorithm as well. The global exploration optimization algorithm can overcome problems such as multimodality, nonlinearity, discontinuity, and non-differentiability. The optimization process adopts algorithms such as population cultivation, survival of the fittest, and random sampling algorithms. It does not involve iteration and can be optimized globally; however, it requires a large amount of calculation.

Appropriate optimization algorithms should be selected to obtain accurate optimization results. The structural optimization process of the pneumatic soft manipulator deformation includes several optimization variables and complex constraints. Furthermore, the implicit iterative algorithm is adopted in the simulation calculation. The objective function presents the characteristics of multimodality, nonlinearity, discontinuity, and non-differentiability, which

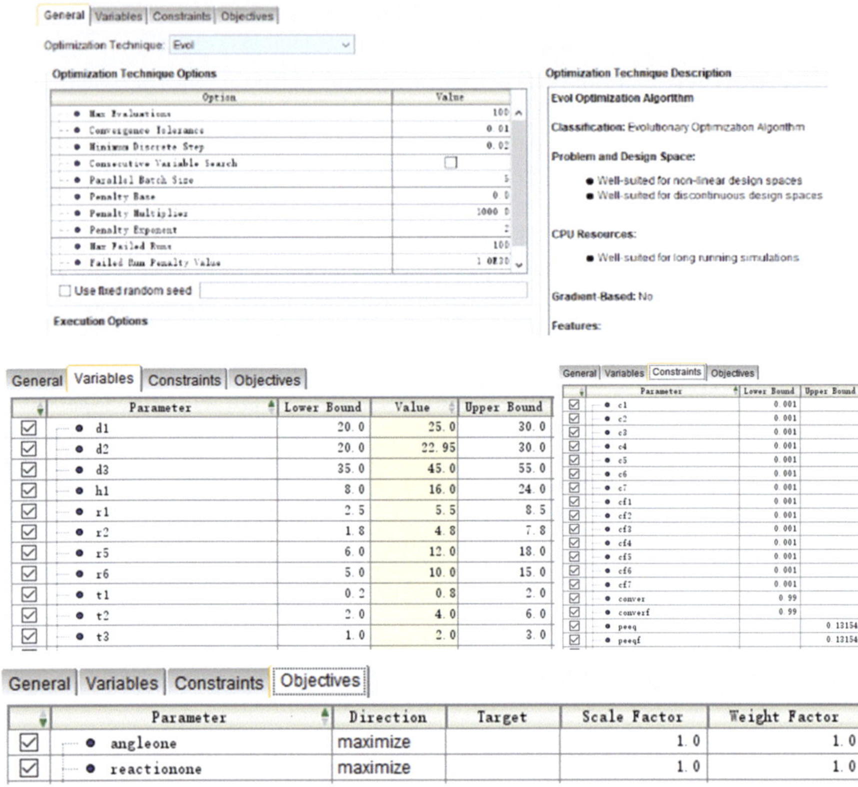

Fig. 3.11 Optimization parameter setting

considerably increase the difficulty of the calculation. Finally, an evolutionary optimization algorithm (Evol) is adopted in the global optimization algorithm [9].

In Evol, a random value of the normal distribution is added to each design variable to realize the mutant design. The expectation and standard deviation of the normal distribution are adaptive and vary during the optimization process. The algorithm can simultaneously change multiple design variables and ensure that there is no duplication of the design points. It can realize parallel computing of multiple design points and select the optimal design point to indicate the next step. It is especially suitable for the optimization process that involves severe nonlinearity, discontinuity, and a large amount of calculation. The optimization parameters are set, including the initial value, upper and lower limits, constraint range, optimization direction, and the target value of the objective function, as shown in Fig. 3.11.

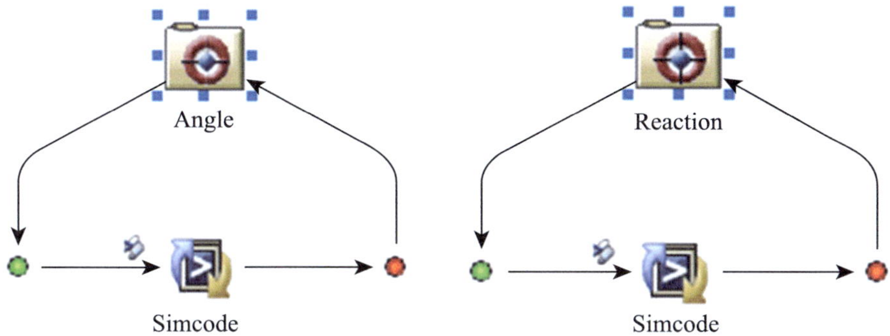

(a) Inclination angle (b) Reaction force

Fig. 3.12 Schematic of single objective optimization of inclination angle and reaction force

3.2.2.2 Optimization Scheme

The structural optimization of the pneumatic soft manipulator involves two load working conditions, as explained before. The optimization objective is to maximize the bending angle and the support reaction force simultaneously. This becomes difficult when multi-objective optimization is involved because a strong coupling correlation exists between different objectives. Therefore, the weighting and normalization method should be adopted to couple multiple objectives into a single objective for optimization. The specific scheme is listed as follows:

(1) The maximum value of the bending angle is obtained as a single optimization objective, Angle_max, the process is shown in Fig. 3.12a.
(2) The maximum value of the reaction force is obtained as a single optimization objective, *F*_max, the process are shown in Fig. 3.12b.
(3) When the bending angle and support reaction force are selected as the multi-objective optimizations, Angle_max and *F*_max are used as the denominators for normalization, respectively, and the target weighting ratio is 1:1, the process are shown in Fig. 3.13.

Three components can be observed in Fig. 3.13: optimization, simulation, and calculator. The optimization module performs optimization design, the Simocode module calls the ABAQUS function and performs the Python parametric program for simulation calculation, and the calculator module performs normalization processing to form a loop and completes the objective optimization.

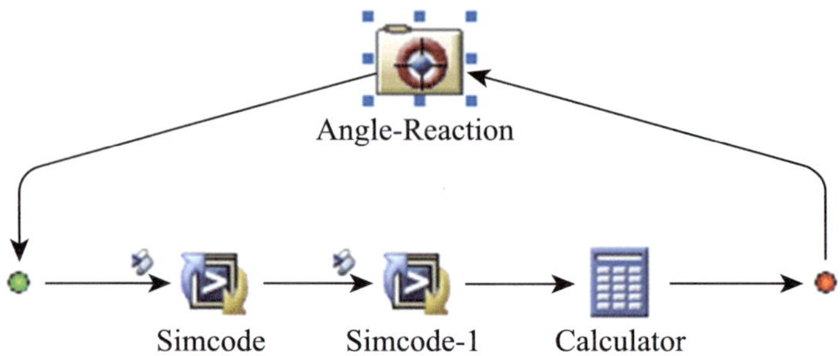

Fig. 3.13 Schematic of multi-objective optimization of inclination angle and support reaction force coupling

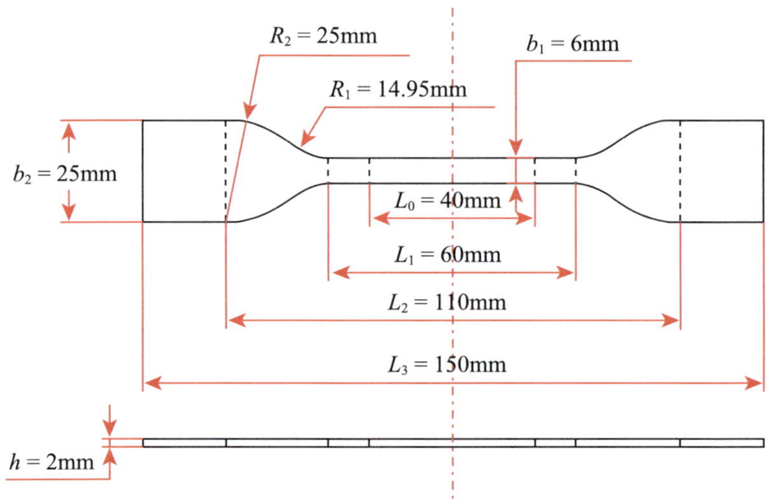

Fig. 3.14 Dumbbell shaped specimen for uniaxial tensile test of nylon material

3.2.3 Determination Method of the Material Constitutive Model

The pneumatic soft manipulator is composed of nylon. The dumbbell-shaped sample shown in Fig. 3.14 is designed for testing elastic-plastic material. Subsequently, the uniaxial tensile test is performed. $L_0 = 40$ mm is the nominal distance, $b_1 = 6$ mm is the specimen width, and $h = 2$ mm is the specimen thickness. The stress-strain data of the interval, L_0, are obtained from the tensile test.

The following assumptions are considered before processing the nominal stress-strain data:

(1) There is no gap inside the object, and the deformation behavior can be described according to continuum mechanics.
(2) The material inside the object presents the same characteristics and satisfies the uniformity hypothesis.
(3) The material properties inside the object are isotropic.
(4) The constitutive relation of the material satisfies Hooke's law in the linear stage and the isotropic hardening model in the plastic stage. The central position of the yield surface remains unchanged, only the shape expands comparably.
(5) The strain in the large deformation behavior is Green strain, and the stress is Cauchy stress.
(6) The deformation of each point within the gauge range in the tensile test is consistent.

The data of the material experiments are often presented in the form of nominal stress and strain. In this case, they should be converted to the true stress and strain. The correlation between the true strain and nominal strain is as follows:

$$\varepsilon = \ln(1 + \varepsilon_{\text{nom}}) \tag{3.1}$$

where, ε represents the true strain or logarithmic strain; ε_{nom} represents the nominal strain.

Considering the incompressibility of plastic deformation and assuming that elastic deformation is incompressible, the correlation between the real stress and nominal stress can be established as follows:

$$\sigma = \sigma_{\text{nom}}(1 + \varepsilon_{\text{nom}}) \tag{3.2}$$

where, σ and σ_{nom} represent the real and nominal stress, respectively.

Figure 3.15 illustrates the nominal (black line) and real (red line) stress-strain curves. The yield strength is calculated in these conditions: the residual strain 0.2%, $\sigma_{0.2} = 15.18$ MPa, $\sigma_{\text{nom}0.2} = 14.90$ MPa, Young's modulus $E = 809.6$ MPa and Poisson's ratio $\nu = 0.35$.

The stress and plastic strain data in the principle of structural finite elements should be definite to facilitate the calculation of the yield function and hardening parameters. The correlation between the different strains is given as follows:

$$\varepsilon_{\text{p}} = \varepsilon - \varepsilon_{\text{e}} = \varepsilon - \frac{\sigma}{E} \tag{3.3}$$

where, ε_{p} and ε_{e} represent the plastic and elastic strains, respectively. Figure 3.16 depicts the true plastic stress-strain curve, where the maximum equivalent plastic strain $\text{peeq}_{\max} = 0.131537$.

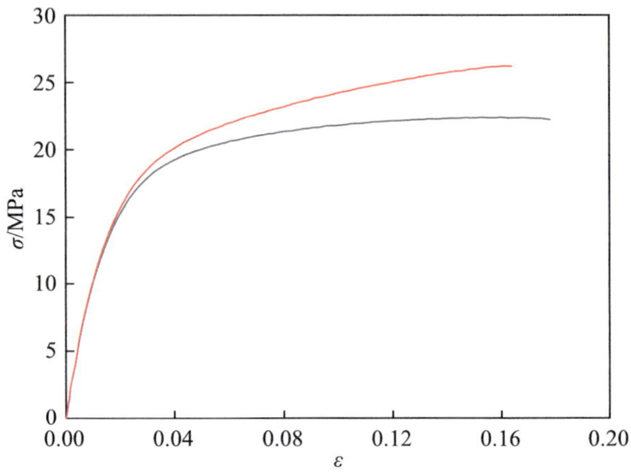

Fig. 3.15 Nominal and real stress-strain curves

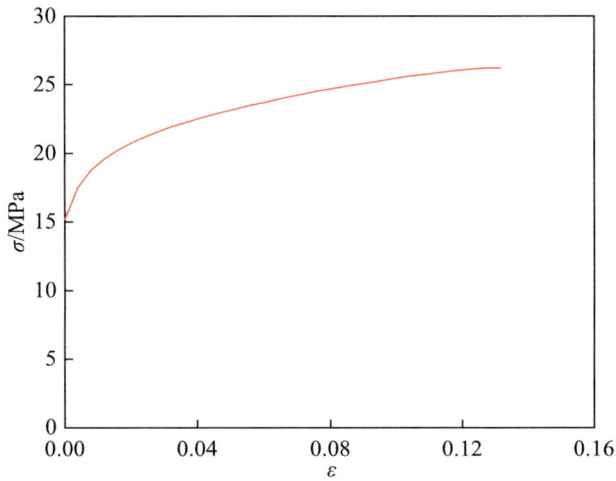

Fig. 3.16 True plastic stress-strain curve

3.3 Sensitivity Analysis of Structural Parameters

The sensitivity of the dimensional parameters should be analyzed to determine the influence of different dimensional parameters on the mechanical properties of the pneumatic soft manipulator. The deformation optimization of the pneumatic soft manipulator includes two load conditions. The effect of the dimensional parameters on the mechanical properties of the pneumatic soft manipulator is analyzed using a single parameter sensitivity analysis method. The analysis process is explained in the following subsections.

Table 3.2 Initial values of design variables and response values

Design variables			Inclination angle/(°)	Reaction force/N
Name	Range/mm	Initial values/mm	Initial response values	
d_1	22.50–27.50	25.00	27.109	111.71
h_1	14.40–17.60	16.00		
t_1	0.72–0.88	0.80		
t_2	3.60–4.40	4.00		
r_1	4.95–6.05	5.50		
r_2	4.32–5.28	4.80		
d_2	20.655–24.245	22.95		
d_3	40.50–49.50	45.00		
r_5	10.80–13.20	12.00		
r_6	9.00–11.00	10.00		
t_3	1.80–2.20	2.00		
n	10–20	16		
Pre/MPa	0.10–0.60	0.30		

3.3.1 Initial Parameter Value and Design Range

Table 3.2 lists the range and initial value of the design variables and the initial response values, and the variables are described in Sect. 3.2.

According to the physical meaning, the variable factors are divided into three categories: structural size, number of airbag sections, and internal pressure. The design range of the structural size is ±10%. The influence of the factors cannot be adequately reflected if the structural size is too small, and interference errors of geometry may occur if it is too large. The number of airbag sections and the internal pressure are independent of the geometric dimensions. The design range can be appropriately expanded to reflect the influence of the factors. Moreover, the resultant number of airbag sections is within 10–20, and the range of the internal pressure lies within 0.10–0.60 MPa.

The design point can be obtained based on the design variable range and the initial value. Figure 3.17 presents the design point of the single-factor analysis and shows that the number of tests for each factor is approximately 11. Each design point has only one factor, which deviates from the initial value when using the control variable method.

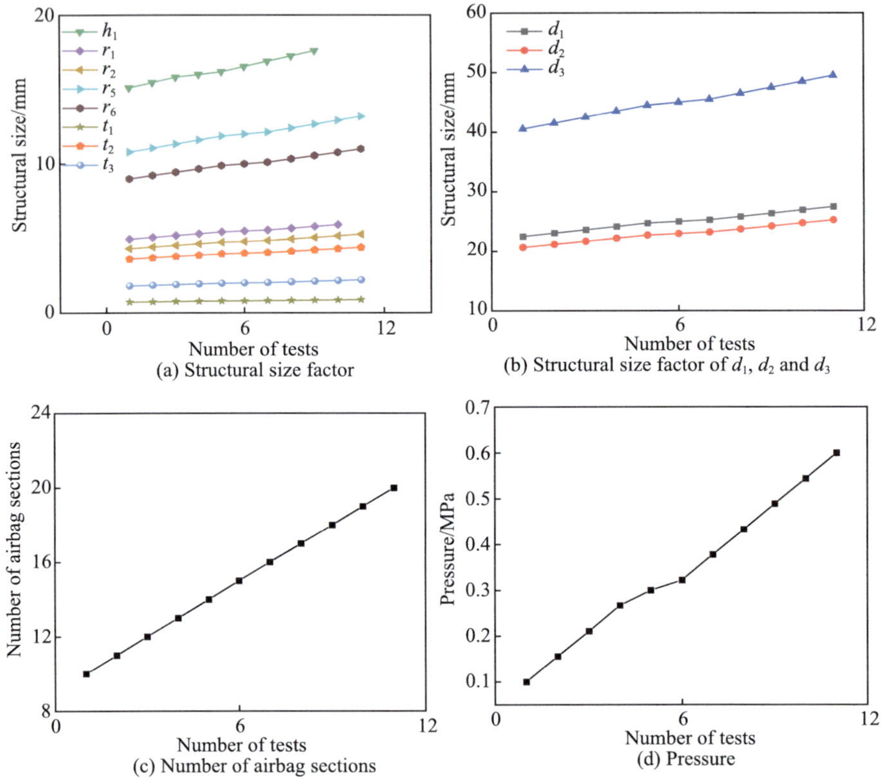

Fig. 3.17 Distribution of experimental design points with different factors

3.3.2 Single Factor Impact Analysis: Inclination Angle

In this load condition, the bottom of the pneumatic soft manipulator is fixed, and one bellow is inflated. Bending deformation is observed, and the bending angle is considered as the response. The results of the single factor impact indicate that some factors have a negligible effect on the response results, whereas some have a small impact on the response results; moreover, there are no rules to follow. The remaining factors significantly affect the response results and are monotonous; further details are presented below.

The change in these factors has a negligible effect on the inclination angle, as shown in Fig. 3.18. The change in one factor cannot significantly change the stiffness in the deformation direction when the univariate factor analysis is performed because the d_2, r_5, r_6, and t_3 factors are related to the thin plate structure, and the direction of the external force and action area are also independent of these factors.

These factors do not significantly affect the response results, as shown in Fig. 3.19, and there are no rules to follow. The change in one factor can affect the structural

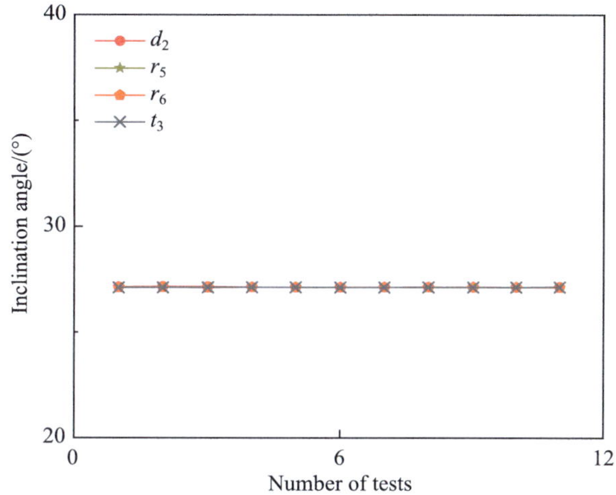

Fig. 3.18 Low sensitivity influence factor of inclination angle

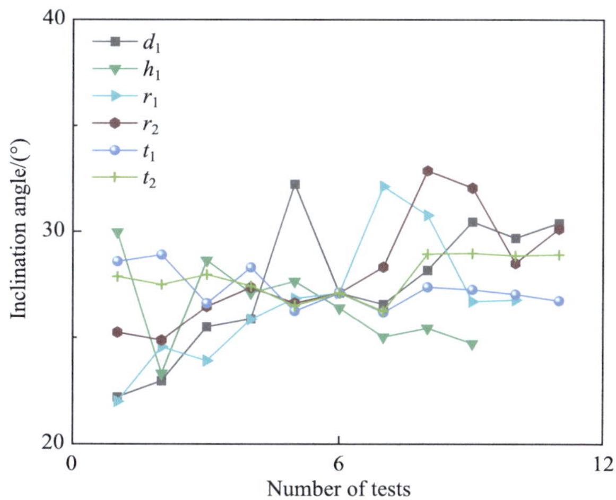

Fig. 3.19 Influence factors of low sensitivity and irregular distribution of inclination angle

stiffness in the deformation direction and change the action area or direction of the external force when performing univariate factor analysis. This is because the factors, d_1, h_1, r_1, r_2, t_1, and t_2, are related to the structure of the bellows, which directly causes a change in the stiffness-displacement-force balance equation. The change in the direction of stiffness and external force may be inconsistent, which leads to irregular displacement response results, as shown in Fig. 3.19.

These factors significantly affect the response results and are monotonic, as shown in Fig. 3.20. The influences of the factors are described as follows:

(1) Parameter d_3 represents the distribution distance between the three airbags. The manipulator inclination angle decreases gradually with the increase in the distribution distance. Essentially, the stiffness reflects the ability to resist the deformation of the structure when the structural design is compact. The stiffness in the

Fig. 3.20 High sensitivity influence factor of inclination angle

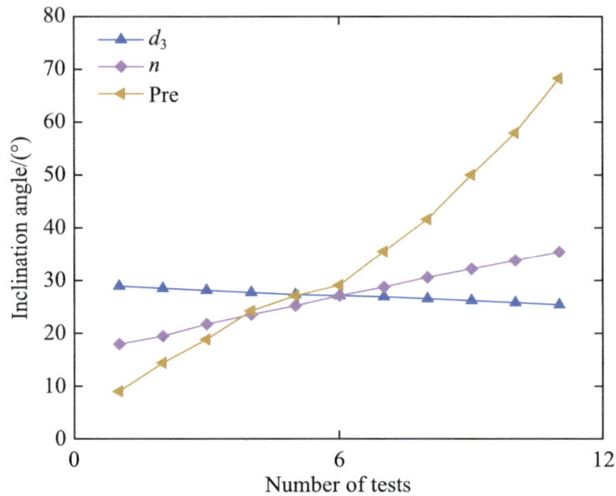

deformation direction is large in the external force and deformation mode. The stiffness mainly reflects the ability to resist the deformation of the local structure when the structural design is scattered. The stiffness in the deformation direction is small in the external force and deformation mode.

(2) Parameter n represents the number of bellow sections. The results demonstrate that the inclination angle of the manipulator gradually increases with an increase in the number of bellow sections. If the bellow is considered as a cantilever beam, the length of the cantilever beam increases when the number of sections increases. The moment of inertia, flexural section coefficient, and stiffness do not change because the structural size remains unchanged. Furthermore, the degree of deformation also changes if only the action position of the external force is changed, even if the magnitude and direction of the external force remain unchanged. Consider a cantilever beam whose free end is subjected to a downward concentrated load. Then, if the length of the cantilever beam increases and the concentrated load at the free end remains unchanged, the deflection at the free end also increases. The force equilibrium state is identical in the two structures, but the moment equilibrium state equation varies; that is, the bending moment increases with an increase in the length of the cantilever beam.

(3) The internal pressure, Pre, is essentially a process parameter. It presents the most significant effect on the response results when compared to the other factors. The results indicate that the inclination angle increases significantly with an increase in the internal pressure. The internal pressure has the most direct impact on the response results because it directly determines the magnitude of the external force.

3.3.3 Univariate Factor Impact Analysis: Reaction Force

Under this load condition, both ends of the pneumatic soft manipulator are simultaneously fixed, and one bellow is inflated to realize internal extrusion. The reaction force of the top end is considered as the response. The load at each position is perpendicular to the action surface, and the top support reaction force reflects the stress condition at the top of the airbag because the internal air pressure load is applied. Despite the entire force balance, a local stress analysis is performed instead of an analysis of the entire stress. The airbag constrained at both ends is simplified as a beam structure, which is internally or locally stressed to produce the extrusion. The effect of the factors on the top reaction force can be described based on an analysis of the effect of the factors on the external force and stiffness (essentially independent of stiffness, only to facilitate analysis). These factors do not significantly affect the reaction force, as shown in Figs. 3.21 and 3.22. The analysis is listed as follows:

(1) The changes in the factors t_1, t_2, r_1, and r_2 of the airbags lead to changes in the local action area of the air pressure when performing univariate analysis. However, these changes present a negligible effect on the transverse section area. Essentially, the section stiffness is not significantly affected by area and shape, and even the changes in the external force and stiffness offset each other. The support reaction force reflects the effect of this influence as an external resultant force on the constraint boundary, as shown in Figs. 3.21 and 3.22.
(2) The parameters d_2, d_3, r_5, r_6, and t_3 are related to the thin plate. Based on the inclination analysis, they do not significantly affect the stiffness and stress area at the loading position in the univariate factor analysis; therefore, they have a minimal effect on the support reaction force.
(3) If the number of airbag sections, n, is compared to the length of the beam, any change in n does not cause a change in the stress and strain; this change in n also does not affect the stability of the structure in the equilibrium state when the external force remains unchanged.

These factors significantly affect the support reaction force and are monotonous, as shown in Fig. 3.23. The influences of the factors are described as follows:

(1) The transverse and longitudinal dimensions of the airbag depend on d_1 and h_1, respectively. The results indicate that the support reaction force increases with an increase in d_1. This is because an increase in d_1 results in an increase in the stress area of the air pressure; thus, the resultant force increases. The reaction force gradually decreases with an increase in h_1. This is because an increase in h_1 indirectly affects the positional correlation of other structures, resulting in the reduction of the transverse area (the longitudinal area will not be affected); thus, the external resultant force decreases.
(2) Similarly, the internal pressure, Pre, is essentially a process parameter, and its influence is the most evident when compared to that of the other factors. The results indicate that the reaction force increases significantly with an increase in the internal pressure.

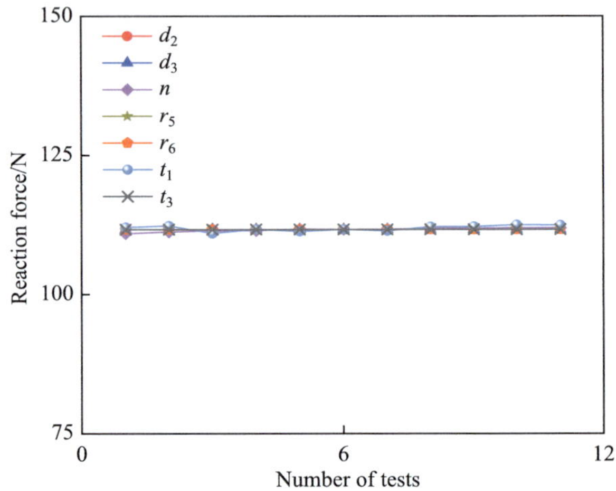

Fig. 3.21 Low sensitivity influence factor of support reaction force (thin plate related)

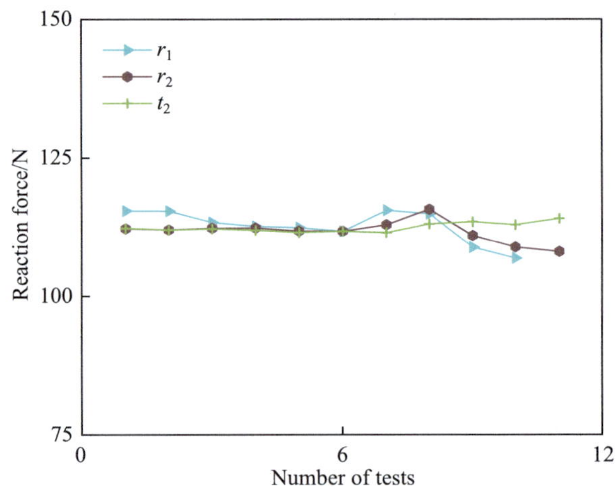

Fig. 3.22 Low sensitivity influence factor of branch reaction force (airbag related)

3.3.4 Analysis Conclusion

From the analysis above, it is concluded that a noticeable monotonic correlation is observed between n, Pre, and the response value, which has a significant impact. In some cases, the "large factors" cover the "small factors" if they are considered as optimization variables, and the optimization results of n and Pre are obtained at the end of the design range. Therefore, only the structural size factor is selected as the optimization variable. Table 3.3 depicts the correlation of the sensitivity of the

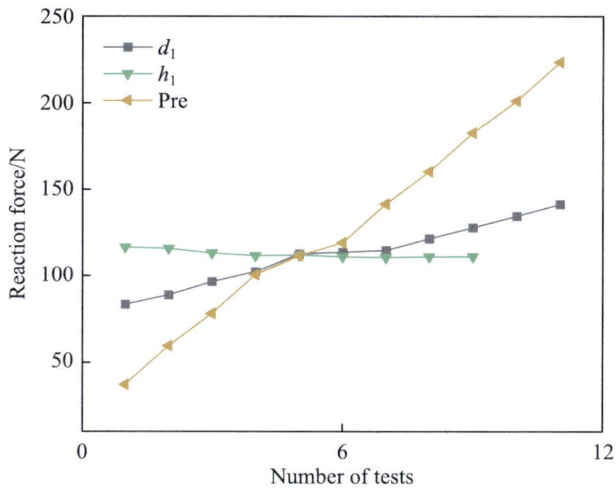

Fig. 3.23 High sensitivity influence factor of branch reaction force

Table 3.3 Sensitivity of structure size factor

Design variables/%	Inclination angle/%	Size factor sensitivity		
		Monotonic trend	Reaction force/%	Monotonic trend
d_1	18.1	Nonmonotonic	25.4	Monotonic increasing
h_1	12.6	Nonmonotonic	2.4	Monotonic decreasing
t_1	4.4	Nonmonotonic	1	Minimum
t_2	4.3	Nonmonotonic	1.1	Nonmonotonic
r_1	19.2	Nonmonotonic	3.8	Nonmonotonic
r_2	13.5	Nonmonotonic	3.4	Nonmonotonic
d_2	1	Minimum	1	Minimum
d_3	6.8	Monotonic decreasing	1	Minimum
r_5	1	Minimum	1	Minimum
r_6	1	Minimum	1	Minimum
t_3	1	Minimum	1	Minimum

structural size factor to the inclination angle and the support reaction force. It can be observed that d_1, h_1, r_1, and r_2 affect the inclination angle by more than 10%, while other factors present a minimal influence. Furthermore, d_1 presents the greatest influence on the reaction force and exhibits a monotonic increasing trend, which is close to 25.4%. The influence of other factors is relatively small.

3.4 Structural Optimization and Performance Evaluation

The sensitive structural size parameters are selected and optimized through three steps. The optimization steps are as follows.

(1) Load case 1 optimization: the inclination angle is considered as a univariate optimization objective to obtain its maximum value.
(2) Load case 2 optimization: the support reaction force is considered as a univariate optimization objective to obtain its maximum value.
(3) Multi-objective coupling optimization: the inclination angle and support reaction force are normalized according to the optimization results of (1) and (2) and weighed for multi-objective coupling optimization.

It is necessary to set the optimization variables, optimization objectives, and constraints before optimization. To ensure the rationality of the structure, it is necessary to set the dimensional constraints. A Boolean value should be set to determine the convergence, and the deformation limit should be set to avoid structural failure. The deformation limit is the maximum equivalent plastic strain, $\text{peeq}_{\text{max}} = 0.131537$.

3.4.1 Establishment of Global Objective Function

The optimization results of the inclination angle and support reaction force are obtained from the univariate optimization objective, and the maximum values of the two optimization objectives are Angle_max and F_max, respectively; these values are then considered as the denominators to normalize the optimization target, as shown below:

$$\text{Opt}(\text{Angle}, F) = f_{\text{Angle}} \cdot \frac{\text{Angle}}{\text{Angle\_max}} + f_F \cdot \frac{F}{F\text{\_max}} \tag{3.4}$$

where, $\text{Angle\_Coefficient} = \dfrac{\text{Angle}}{\text{Angle\_max}}$ represents the inclination angle normalization coefficient and $F\text{\_Coefficient} = \dfrac{F}{F\text{\_max}}$ represents the normalization coefficient of the reaction force, while temporarily considering $f_{\text{Angle}} : f_F = 1:1$.

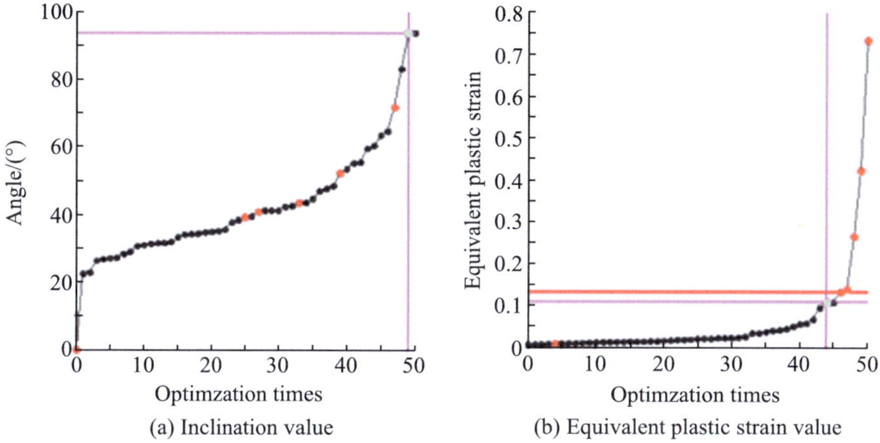

Fig. 3.24 Inclination angle optimization results of the pneumatic manipulator

3.4.2 Calculation of the Maximum Value of the Univariate Objective

3.4.2.1 Inclination Angle Optimization

The optimization times are set to 100, but the number of successful calculations is approximately 50 owing to structural errors and convergence. Figure 3.24a presents the numerical statistical results of inclination angle optimization, where the red point indicates the failure point, the black point indicates the non-optimal point, and the green point indicates the optimal point. Figure 3.24b presents the global maximum equivalent plastic strain of the simulation model. It can be observed that the equivalent plastic strain calculated at some design points exceeds the maximum plastic strain, and these points are finally determined as the failure point.

Table 3.4 lists the design variables and optimization results. The final value, Angle_Max = 93.787° is observed to be thrice the initial value of 27.109°.

3.4.2.2 Optimization of Reaction Force

Figure 3.25a presents the numerical statistical results of the optimization of reaction force, where red indicates the failure point, black indicates the non-optimal point, and green indicates the optimal point. Figure 3.25b presents the global maximum equivalent plastic strain of the simulation model.

Table 3.5 lists the design variables and optimization results, and the final value of F_max = 179.2N, is observed to be 1.6 times the initial value of 111.71N.

Table 3.4 Inclination angel design variables and optimization results

Name	Design variables/mm			Inclination angle/(°)	
	Range	Initial value	Results	Initial value	Results
d_1	20.00–30.00	25.00	28.4	27.109	93.787
h_1	8.00–24.00	16.00	18.88		
t_1	0.20–2.00	0.80	0.512		
t_2	2.00–6.00	4.00	4.4		
r_1	2.50–8.50	5.50	6.1		
r_2	1.80–7.80	4.80	6.0		
d_2	20.00–30.00	22.95	21.55		
d_3	35.00–55.00	45.00	47.4		
r_5	6.00–18.00	12.00	15.6		
r_6	5.00–15.00	10.00	11.4		
t_3	1.00–3.00	2.00	1.68		

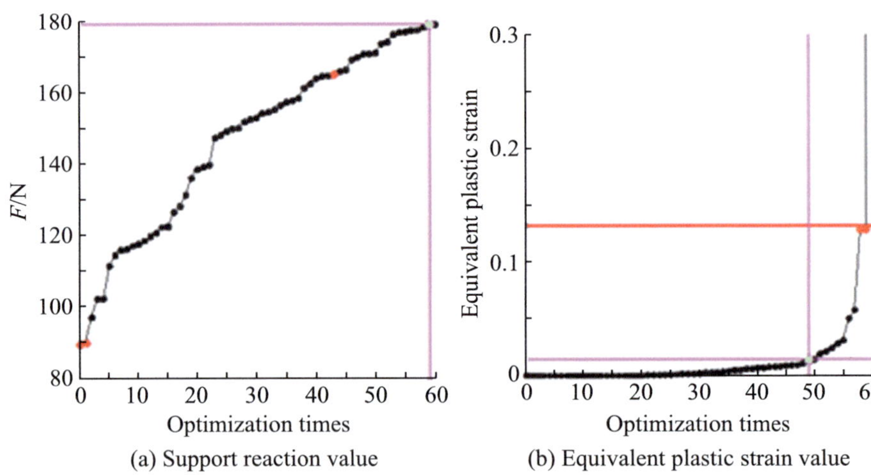

Fig. 3.25 Optimization results of the support reaction force of the pneumatic manipulator

3.4.3 Implementation of Global Optimization

The maximum values obtained from the above univariate optimization results, according to Eq.(3.4), are Angle_max = 93.787° and F_max = 179.2N, and the optimization times are set to 100. These values are the statistical results of Angle_coefficient, where red indicates the failure point, black the non-optimal point, blue the maximum value of an optimization variable at this time, and green the optimal point, as shown in Fig. 3.26a. Figure 3.26b presents the statistical results of F_Coefficient.

Table 3.5 Design variables and optimization results of support reaction force

Name	Design variables/mm			Reaction force/N	
	Range	Initial value	Results	Initial value	Results
d_1	20.00–30.00	25.00	30.0	111.71	179.2
h_1	8.00–24.00	16.00	13.44		
t_1	0.20–2.00	0.80	0.476		
t_2	2.00–6.00	4.00	5.28		
r_1	2.50–8.50	5.50	3.7		
r_2	1.80–7.80	4.80	5.04		
d_2	20.00–30.00	22.95	22.35		
d_3	35.00–55.00	45.00	51.4		
r_5	6.00–18.00	12.00	18.0		
r_6	5.00–15.00	10.00	9.4		
t_3	1.00–3.00	2.00	2.08		

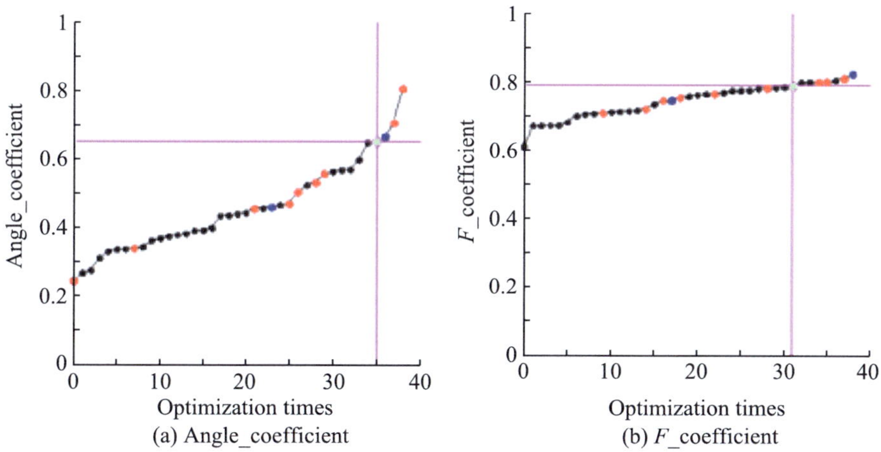

Fig. 3.26 Normalization coefficient statistics

When the coupling optimization objective obtains the optimal solution, the normalization coefficient of the inclination angle is Angle_Coefficient = 0.65252, and the normalization coefficient of the support reaction force is *F*_Coefficient = 0.78875. From these values, Angle = 64.054° and F = 144.105N. Table 3.6 lists the results of the optimization variables and optimization objectives.

Table 3.6 Optimization results of coupling of inclination angle and reaction force

Name	Design variables/mm			Inclination angle/(°)	Reaction force/N
	Range	Initial value	Results	Initial value/results	Initial value/results
d_1	20.00–30.00	25.00	28.0	27.109/64.054	111.71/144.105
h_1	8.00–24.00	16.00	15.68		
t_1	0.20–2.00	0.80	0.512		
t_2	2.00–6.00	4.00	3.76		
r_1	2.50–8.50	5.50	4.66		
r_2	1.80–7.80	4.80	6.0		
d_2	20.00–30.00	22.95	22.95		
d_3	35.00–55.00	45.00	47.0		
r_5	6.00–18.00	12.00	9.36		
r_6	5.00–15.00	10.00	9.0		
t_3	1.00–3.00	2.00	1.88		

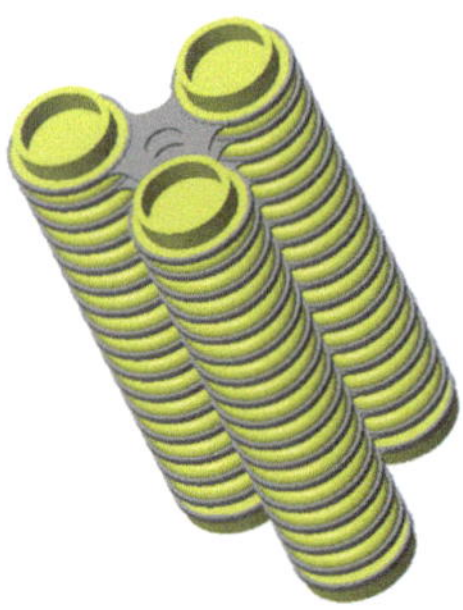

Fig. 3.27 Optimization result pneumatic manipulator

3.4.4 Optimization Results of Pneumatic Manipulator

The pneumatic soft manipulator is redesigned according to the optimized results, as shown in Fig. 3.27.

The optimized pneumatic manipulator is simulated and analyzed under two load conditions. Figure 3.28 presents the simulation results.

It can be observed that under load condition 1, the manipulator bends, and plastic deformation occurs owing to the action of internal pressure. The equivalent plastic strain is less than the maximum equivalent plastic strain, $peeq_{max} = 0.131537$. Furthermore, the residual strain might exist after removing the internal air pressure because the plastic deformation cannot be recovered. Therefore, plastic deformation should preferably not occur under the action of the internal air pressure (the air pressure in this simulation is 0.3 MPa). The von Mises stress and peeq strain generated by the action of the internal air pressure under load condition 2 are less than those of load condition 1, and almost no plastic strain occurs in load condition 2, which lies

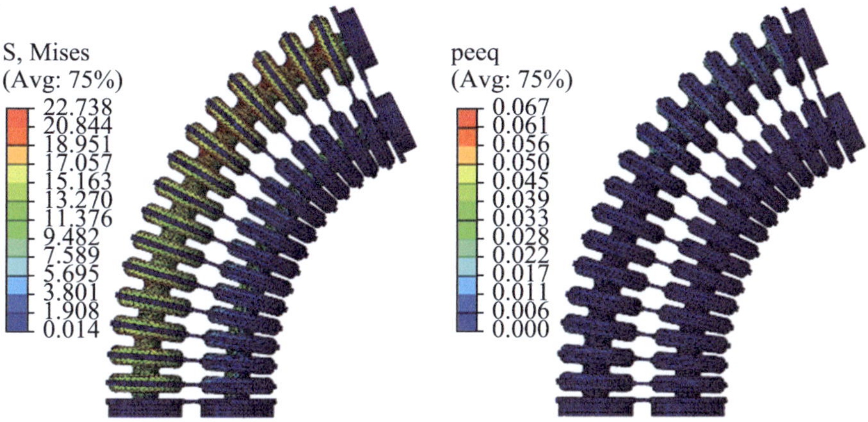

(a) Working condition 1: the bottom end is fixed and the manipulator is free to bend (the von Mises equivalent stress (Pa) and equivalent plastic strain)

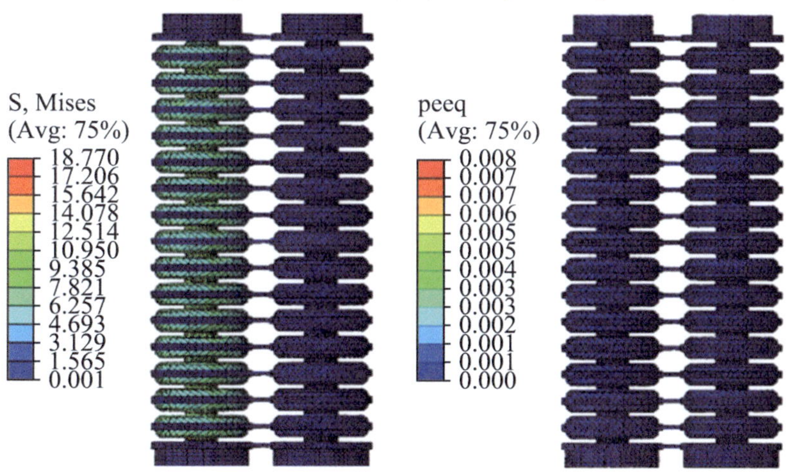

(b) Working condition 2: both ends are fixed, and the manipulator is extruded inward (the von Mises equivalent stress (Pa) and equivalent plastic strain)

Fig. 3.28 Simulation analysis diagram of the optimized pneumatic manipulator

within the safe deformation range. The simulation results show that the deformation of the optimized pneumatic soft manipulator under the given internal pressure is within the safe range and meets the design requirements. This study provides a new optimization method and scheme for the structural optimization of bellows, which greatly increases the mechanical properties and load capacity of the pneumatic soft manipulator, and lays a foundation for the application of the soft manipulator in more complex task scenarios in the future.

3.5 Summary

This study proposes a structural analysis and optimization design method for a pneumatic soft manipulator, considering the bending inclination angle and support reaction force as the evaluation objectives. The main conclusions of this study are as follows:

(1) The geometric parameterization program was developed independently. The nonlinear algorithm adopted in the simulation calculation was coupled with the Evol optimization algorithm, and the characteristic parameter optimization was realized efficiently and stably. Two types of constraint relations between each geometric feature of the pneumatic soft manipulator were established. Finally, a sensitivity analysis of each parameter was conducted to determine high-sensitivity parameters.
(2) Global objective optimization of the inclination angle and reaction force was performed using the normalization and weighting strategy combined with the Evol optimization algorithm. The support reaction force and inclination angle increased by 29% and 132%, respectively, and the optimal parameter combination was obtained.
(3) Geometric errors such as interference may occur when the geometric topology changes or the structure significantly changes because the pneumatic soft manipulator had many coupling optimization parameters. Topology optimization can be used for pneumatic soft manipulator design in the future to obtain the optimal design with wider applicability.

In terms of sensitivity analysis, the use of multi-factor analysis methods such as orthogonal tests, central factor combination tests, and Latin hypercube tests leads to a large amount of calculation. In addition, the results do not easily converge within a reasonable range. Genetic algorithms or artificial neural networks should be used to further analyze the existing characteristic parameters in the future, and the inclination angle and support reaction force of the pneumatic soft manipulator can be further developed when combined with topology optimization. This research greatly increases the mechanical properties and load capacity of the pneumatic soft manipulator, and widely increase the application fields for pneumatic soft manipulator such as ground rescue, man-machine interaction and so on.

References

1. Hu W, Mutlu R, Li W et al (2018) A structural optimisation method for a soft pneumatic actuator. Robotics 7(2):24
2. Liu CH, Chen LJ, Chi JC et al (2022) Topology optimization design and experiment of a soft pneumatic bending actuator for grasping applications. IEEE Robot Autom Lett 7(2):2086–2093
3. Chen F, Song Z, Chen S et al (2023) Morphological design for pneumatic soft actuators and robots with desired deformation behavior. IEEE Trans Rob 39(6):4408–4428

4. Bieze TM, Largilliere F, Kruszewski A et al (2018) Finite element method-based kinematics and closed-loop control of soft, continuum manipulators. Soft Rob 5(3):348–364
5. Largilliere F, Verona V, Coevoet E et al (2015) Real-time control of soft-robots using asynchronous finite element modeling. IEEE international conference on robotics and automation (ICRA). Seattle, WA, USA, pp 2550–2555
6. Lee TS, Lin YJ (1993) Shear deformation effect in design considerations of flexible manipulators. Robotica 11(1):83–92
7. Ypma TJ (1995) Historical development of the Newton-Raphson method. SIAM Rev 37(4):531–551
8. Wang W, Mo R, Zhang Y (2011) Multi-objective aerodynamic optimization design method of compressor rotor based on ISIGHT. Proc Eng 15:3699–3703
9. Bäck T, Schwefel HP (1993) An overview of evolutionary algorithms for parameter optimization. Evol Comput 1(1):1–23

Chapter 4
Experimental Validation of a Composite Soft Actuator

The pneumatic soft actuator is the basic module of the pneumatic soft manipulator. The composite pneumatic actuator can be fabricated by selective laser sintering (SLS) 3D printing technology. However, there is a high risk of fracture in such a composite structure when its inner pressure increases to a high level. In order to make sure the safety of the pneumatic actuator, its reliability is necessary to be evaluated. In this chapter, the reliability and failure of a composite pneumatic actuator are analyzed and tested. In Sect. 4.1, the FEM model of the pneumatic actuator is developed in ABAQUS. The pneumatic experiment system is designed and developed. The shape deformations obtained from the theoretical model, FEM simulation and experiment are compared. In Sect. 4.2, the pressure bearing experiments are conducted. The pulling force of the pneumatic actuator is tested. Finally, the conclusions are presented in Sect. 4.3.

4.1 Shape Deformation Simulation and Experiment

4.1.1 FEM Simulation in ABAQUS

The design of the pneumatic actuator is conducted in SolidWorks [1], as shown in Fig. 4.1. The IGS file of the design is exported from SolidWorks, and then imported into ABAQUS [2]. The components must be merged first into a single part after importing. Next, the materials are defined using Elastic-Isotropic. The pressure is set on the inner surface of one pneumatic bellow. The type of self-contact is defined on the outer surface of the other two pneumatic bellows. The bottom of the actuator is set with a fixed boundary condition. The pneumatic actuator is meshed using totally 420055 lincar tctrahcdral clcment of type C3D4H, as shown in Fig. 4.1.

X. Zhang et al., *Design and Implementation of Soft Robotic Manipulators*,
https://doi.org/10.1007/978-981-95-1657-5_4

Fig. 4.1 FEM model of the composite pneumatic actuator

4.1.2 Design of Pneumatic Experiment System

In order to validate its correctness, a prototype of the composite pneumatic actuator is fabricated by selective laser sintering (SLS) 3D printing technology [3] with nylon composite powder. The pneumatic control system of the actuator is designed and developed, as shown in Fig. 4.2.

In Fig. 4.2a, it is seen that the pneumatic control system includes a composite actuator, a FESTO motion terminal VETM, a BECKHOFF PLC controller, a computer, an air supply, electrical wires, and air pipeline. VTEM is able to control eight pipelines at a proportional pressure with four piezo pilot valves integrated in a valve terminal. The CX9020 BECKHOFF PLC controller is used to offer multi-pipeline air pressure control for the VTEM. The experiment system is shown in Fig. 4.2b. The shape deformation of the composite actuator changing with the increasing of the air pressure and its fracture situations are studied based on this system.

4.1.3 Experimental Validation

Based on the FEM model in ABAQUS, the shape deformation and stress analysis can be obtained, as shown in Fig. 4.3a. In Fig. 4.3a, one of the three bellows is inflated with gas pressures from 1.5bar to 3.0bar, 0.5bar as the interval. Simultaneously, the pneumatic experiments with the same pressure conditions are also conducted. The experimental results are shown in Fig. 4.3b. The other parameters of the pneumatic actuator are listed in Table 4.1.

In Table 4.1, D is the diameter of the maximum cross section of the actuator; d_r is the diameter of the minimum cross section of the actuator; t_1 is the thickness of the bellow shell; L_0 is the length of the small bellow unit shown in Fig. 4.2b; n is the number of the bellow unit; E is the elastic modulus of the bellow material, which is made of nylon; d_2 is the distance between the centers of two bellows; t_2 is the

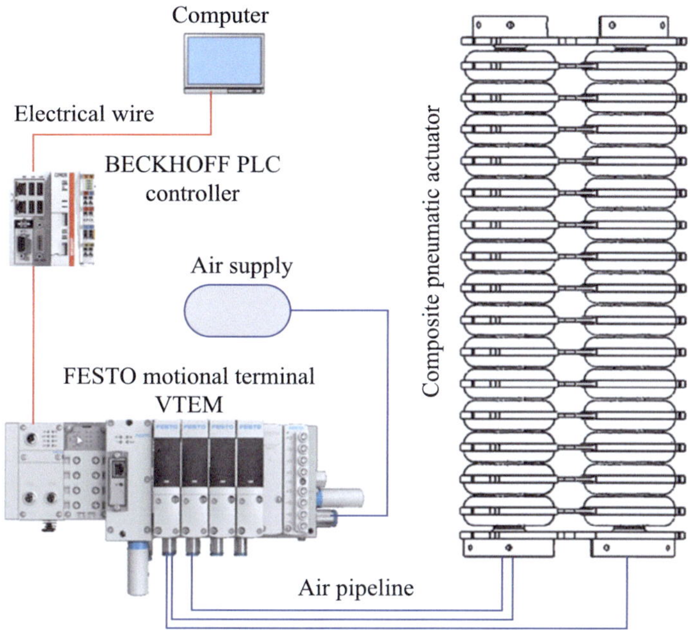

(a) Pneumatic control system

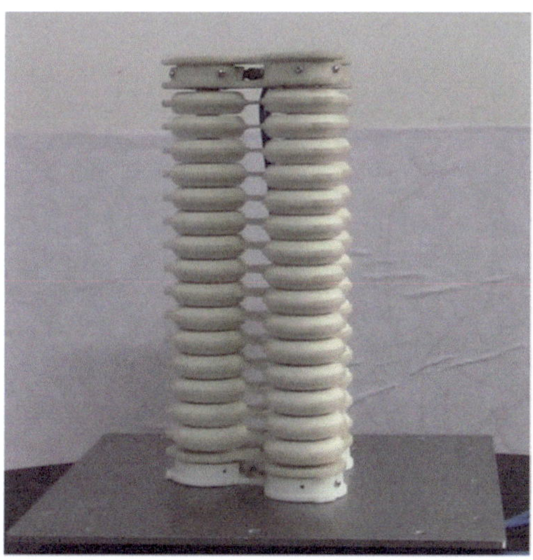

(b) Experiment system

Fig. 4.2 Pneumatic control system of the actuator

thickness of the rigid plate connecting the bellows. The results comparison between the FEM simulation and the experiment are shown in Fig. 4.3.

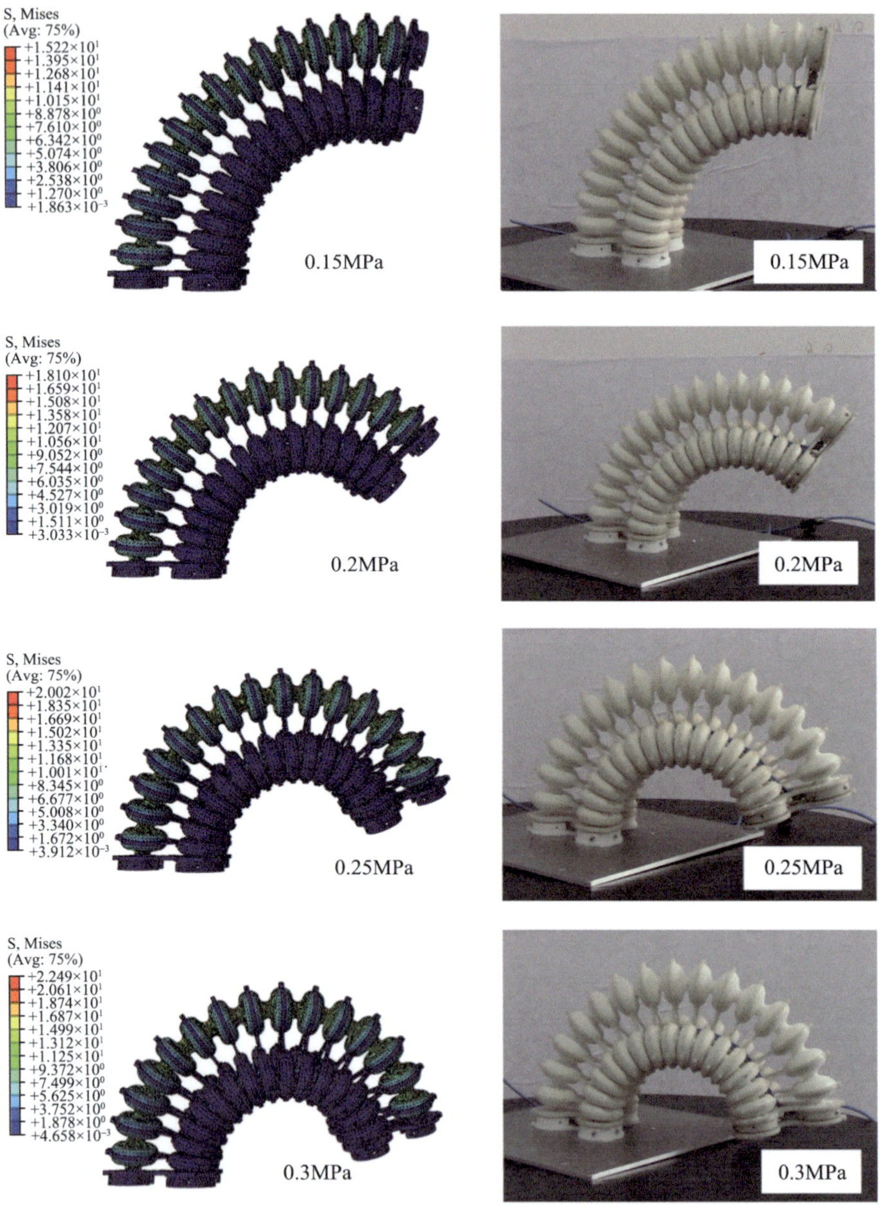

(a) Simulation results (b) Experiment results

Fig. 4.3 Comparison between ABAQUS and experiment results

From Fig. 4.3, it can be seen that the shape deformations obtained from the FEM model and the experiment show good agreement. Thus, the FEM modelling and analysis in ABAQUS have been validated by the experiment.

Table 4.1 The parameters of the pneumatic actuator

Parameter	D	d_r	t_1	L_0
Value	0.05m	0.018m	0.0012m	0.018m
Parameter	n	E	d_2	t_2
Value	15	125 MPa	0.0078m	0.002m

4.2 Implementation Analysis of the Pneumatic Actuator

4.2.1 Tensile Damage Analysis

The maximum tensile stress increases rapidly with the increase of the air pressure [4]. It can be predicted that the tensile damage [5] will happen when the air pressure increases to a certain value. In order to verify this prediction, the pressure bearing experiments of the three bellows are conducted respectively. In each case, only one bellow is inflated with an increasing air pressure. The tensile damage happens in the first bellow when the air pressure increases to 0.31 MPa and keeps a few seconds, as is shown in Fig. 4.4a. The tensile damage also happens in the second bellow when the pressure increases to 0.32 MPa and keeps a few seconds, as is shown in Fig. 4.4b. Nevertheless, the blasting damage happens in the third bellow when the pressure increases to 0.24 MPa, as is shown in Fig. 4.4c.

In Fig. 4.4a and b, it can be seen that the fracture sections of the two bellows are the typical tensile damage section. The sites of the damage sections are both the connection between the two arc surfaces, as is shown in Fig. 4.5. The damage pressures are respectively 0.31 MPa and 0.32 MPa. This kind of damage performance validates the prediction of the proposed model and the FEM simulation. In Fig. 4.4c, the blasting damage happens at a pressure of only 0.24 MPa. It is because there is a small weakness point forming in the third bellow during the 3D printing process. It indicates that the stability of the current 3D printing technology [6] in fabricating thin-shell structures still needs to be improved.

From Fig. 4.4 and 4.5, it can be concluded that the normal fracture pattern of the composite bellow is the tensile damage. The maximum tensile pressure is about 0.3 MPa, and the maximum tensile stress is below 0.3 MPa. Considering the safety and the instability of the 3D printing technology, the working pressure of the pneumatic actuator should be controlled at 0.15 MPa. In order to increase the pressure bearing capacity, the connection between the two arc surfaces should be strengthened. Moreover, if we want to increase the working pressure of the composite pneumatic actuator, the parameters should be designed and optimized based on the proposed model or the FEM simulation results.

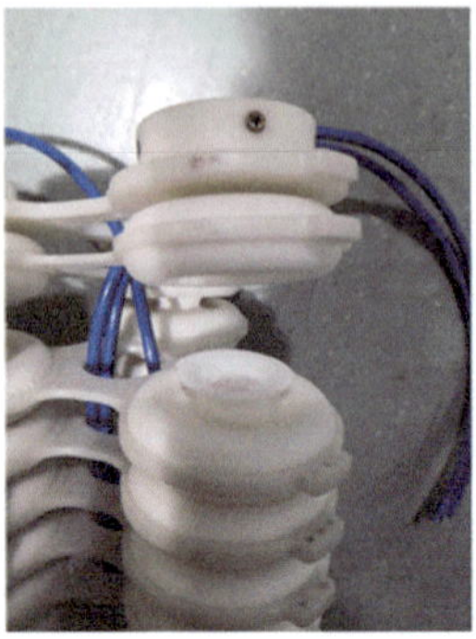

(a) Tensile damage of the first bellow (0.31MPa)

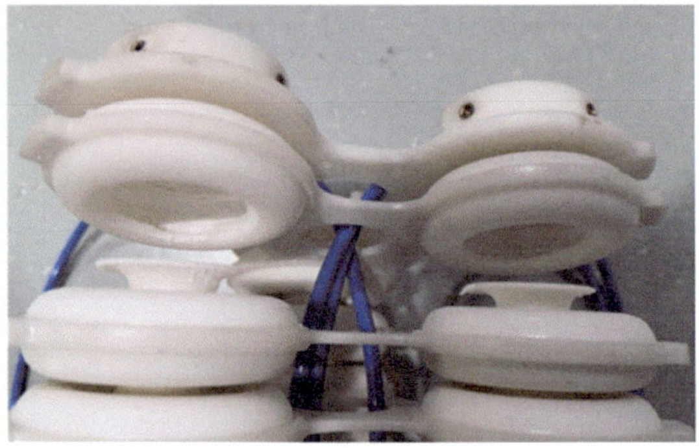

(b) Tensile damage of the second bellow (0.32MPa)

(c) Blasting damage of the third bellow(0.24MPa)

Fig. 4.4 Fracture section of the composite pneumatic actuator

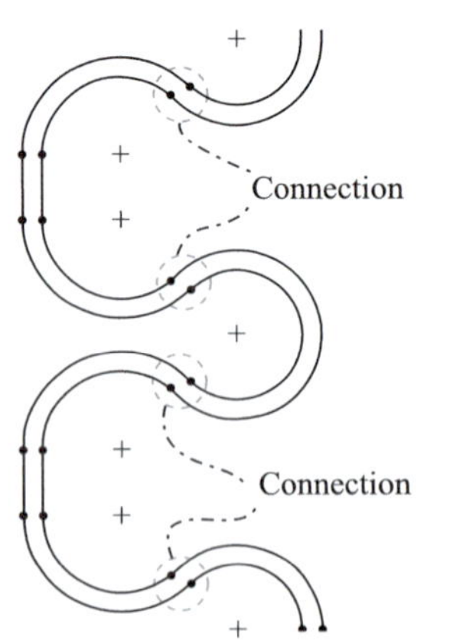

Fig. 4.5 The connection between the two arc surfaces

4.2.2 Pulling Force Analysis

The pneumatic actuator is made up of three inflatable bellows, which are connected and fixed by a series of thin rigid plates. The bellows will elongate in response to the pressured gas. The movement and constraints between the bellows and the thin rigid plates lead to the constrained orientation of the pneumatic actuator. Due to the fact that the pneumatic actuator is to provide the directional actuation force for the soft manipulator, the pulling force under different gas pressure is its key performance [7]. Thus, the experiment to reveal the relationship between the pulling force and the inputting pressure is designed and developed, shown in Fig. 4.6.

In Fig. 4.6, the pneumatic actuator is suspended. The pressured gas is inputted from the top side. The tension meter is connected with the bottom of the actuator using the steel rope. Thus, the pulling force of the pneumatic actuator can be measured by the tension meter when the pressured gas is inputted into the relative bellow. The bending actuation of the pneumatic actuator has two patterns: actuation by inputting pressured gas in double bellows and actuation by inputting pressured gas in single bellow. The pulling forces versus the inputting pressures in these two patterns are recorded in the following Table 4.2.

In Table 4.2, it is seen that the gas pressure changes from 0.2 to 0.5 bar, the pulling force of the double bellows pattern is from 5.1 to 11.1 N, while the pulling force of the single bellow pattern is from 6 to 10.9 N. The fitting results of the pulling force versus inputting pressure are shown in Fig. 4.7.

From the distribution of the test data shown in Fig. 4.7, it can be seen that the pulling force versus the gas pressure is almost linear. Therefore, the linear polynomial fitting equation is adopted [8]. The fitting result in the double bellows pattern is:

$$F_{\text{pulling}} = 19.6117 P_{\text{gas}} + 1.1908 \tag{4.1}$$

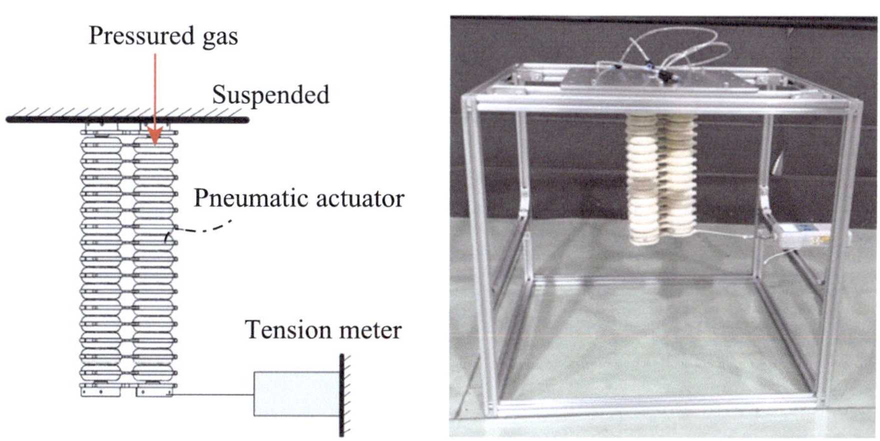

Fig. 4.6 Test system of the pneumatic actuator

Table 4.2 Pulling force versus inputting pressure

Input pressured gas in double bellows		Input pressured gas in single bellow	
Gas pressure/bar	Pulling force/N	Gas pressure/bar	Pulling force/N
0.20	5.10	0.20	6.00
0.25	6.20	0.25	6.80
0.30	7.10	0.30	7.70
0.31	7.20	0.35	8.50
0.40	8.90	0.40	9.30
0.45	10.00	0.45	10.10
0.50	11.10	0.50	10.90

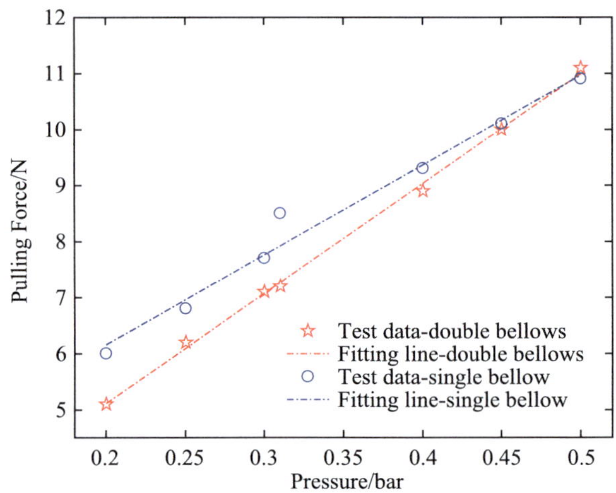

Fig. 4.7 Fitting results of the pulling force versus inputting pressure

The fitting result in the single bellow pattern is:

$$F_{\text{pulling}} = 16.0268 P_{\text{gas}} + 2.9536 \tag{4.2}$$

From Eqs. (4.1) and (4.2), it is known that the gradient of the fitting line in the double bellows pattern is 19.6117, a little higher than the gradient of the fitting line in the single bellow pattern (16.0268). In the condition that the pulling force in these two patterns is equal, the gas pressure is about 0.492bar. It means that the double-bellows actuation is stronger than the single-bellow actuation when the gas pressure is more than 0.492bar.

4.3 Summary

In this chapter, the reliability of the composite pneumatic actuator was studied by FEM simulation and experiment. The FEM simulation of the actuator was conducted by using ABAQUS software. The prototype of the composite pneumatic actuator as well as its pneumatic control system was designed and developed. Through these methods, the shape deformation and the stress calculated were validated by the experiment and the FEM simulation. Moreover, the bearing pressure of the composite pneumatic actuator was about 0.31 MPa, and the maximum tensile stress was below 30 MPa. The fracture sections of the bellows are the typical tensile damage section, and the connections between the two arc surfaces are the vulnerable spots to be damaged. Due to the instability of the 3D printing technology, there is still blasting risk on the pneumatic bellows. The safe work pressure of the composite pneumatic actuator should be controlled at 0.15 MPa. The pulling force of the pneumatic actuator determines the movement ability of the soft manipulator. According to the test results, the pulling force of the pneumatic actuator is stronger in the double-bellows pattern when the gas pressure increases to a certain level. These investigations and conclusions presented in this chapter provide an important and helpful guide to improve the reliability of the composite pneumatic actuator.

References

1. Matsson J (2023) An introduction to solid works flow simulation 2023. SDC Publications
2. Barbero EJ (2023) Finite element analysis of composite materials using Abaqus. CRC Press, Boca Raton
3. Charoo NA, Barakh Ali SF, Mohamed EM et al (2020) Selective laser sintering 3D printing–an overview of the technology and pharmaceutical applications. Drug Dev Ind Pharm 46(6):869–877
4. Martinez RV, Glavan AC, Keplinger C et al (2014) Soft actuators and robots that are resistant to mechanical damage. Adv Func Mater 24(20):3003–3010
5. Marechal L, Balland P, Lindenroth L et al (2021) Toward a common framework and database of materials for soft robotics. Soft Rob 8(3):284–297
6. Wallin TJ, Pikul J, Shepherd RF (2018) 3D printing of soft robotic systems. Nat Rev Mater 3(6):84–100
7. Li Y, Chen Y, Ren T et al (2018) Precharged pneumatic soft actuators and their applications to untethered soft robots. Soft Rob 5(5):567–575
8. Palomares E, Nieto AJ, Morales AL et al (2017) Dynamic behaviour of pneumatic linear actuators. Mechatronics 45:37–48

Chapter 5
Improved Design of the Variable Cross-Section for the Trunk-Like Soft Manipulator

Due to their high motion flexibility and outstanding environmental adaptability, inflatable soft drivers have gradually been adopted to the fields of medical treatment, home service and human movement assistance. They have shown advantages of flexible operation and soft contact. In this chapter, an optimization method for inverse kinematics solving of the inflatable soft driver is proposed, which realizes the fast modeling and operation control. In general, the input air pressure of the inflatable driver forms an n-dimensional vector $\boldsymbol{p}$ and the output is an m-dimensional vector $\boldsymbol{r}$ formed by the position of the end of the driver. The forward kinematics model can be described as the calculation of the end position with the given input pressure. The inverse model can be described as calculation of the input pressure with the given end position, i.e., the inverse kinematics problem. At present, the forward models usually adopt the precise or approximate method based on geometric deformation or finite element simulation and so on. In contrast, the inverse kinematics model is usually not unique due to the redundancy of degrees of freedom, which brings great challenges to solve the inverse kinematics problem. Existing inverse kinematics solving methods, such as the numerical solution of the Jacobian matrix which is not non-square matrix or machine learning methods are often faced with numerical instability or too many training samples, especially when dealing with problems with high redundancy dimensions and restricted boundaries. Aimed at this problem, this chapter presents an optimization method for solving the inverse kinematics problem of the inflatable soft driver. We approximate the relationship between the pressure input and the end position output of the inflatable soft driver into the form of a fitting function with the specified order, in which the order of the fitting model can be dynamically adjusted according to the precision requirements and the number of sampling points. Based on the forward fitting model, the inverse kinematics problem can be abstracted into an optimization problem with constraints. Specifically, it can be described as follows: the designed variables are the air pressure of each cavity of the soft manipulator and the constraint condition is the upper pressure limit of each cavity and other geometric boundary conditions; the objective function is to minimize the deviation between the end of the driver and the specified target position.

X. Zhang et al., *Design and Implementation of Soft Robotic Manipulators*,
https://doi.org/10.1007/978-981-95-1657-5_5

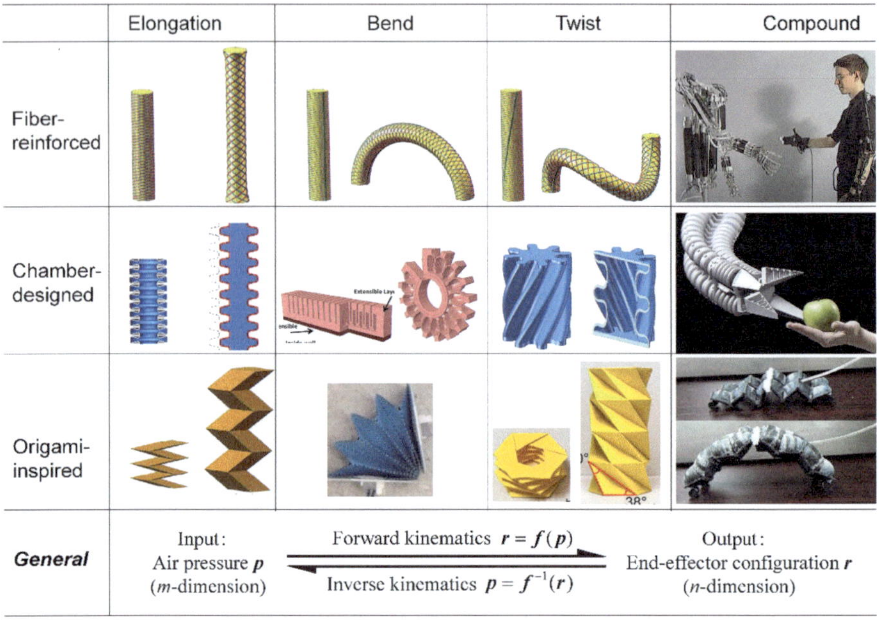

Fig. 5.1 Forward kinematics and inverse kinematics of different inflatable soft drivers

Therefore, the end position of the driver can be located to the specified target by optimization solving based on the inverse kinematics model. It improves the robustness of the control system while ensuring the convergence of the motion. Furthermore, the optimization method can be used as a basic part of closed-loop feedback control to achieve accurate control of inflatable soft drivers.

The n-dimensional vector of inputting air pressures $\boldsymbol{p}$, and the m-dimensional position vector of the effector $\boldsymbol{r}$. The forward kinematic model can be expressed as $\boldsymbol{r} = f(\boldsymbol{p})$ and the inverse kinematic model can be written as $\boldsymbol{p} = f^{-1}(\boldsymbol{r})$. Various methods of building forward models have been as shown in Fig. 5.1, including analytical method based on geometric deformation assumptions, FEM simulations, as well as the general fitting model proposed in this chapter. However, the reverse solution of the redundant model ($n > m$) transforming from the external movement coordinates $\boldsymbol{r}$ to the intrinsic inputting inverse commands $\boldsymbol{p}$ has still remained challenging. This is largely due to the fact that the solution is usually non-unique because the redundant actuator has more degrees of freedom than necessary to solve the task.

Various explorations have been made on solving the inverse problem including analytical solutions, numerical approaches, and machine learning methods. For the models with high redundant dimensionality, restrictive boundaries and other criteria, it is virtually difficult to establish an analytical solution, which is more suitable for low-dimensional manipulators such as rigid manipulators other than high-dimensional continuum actuators such as the multi-chamber pneumatic actuator mentioned above. Although numerical approaches in the literature solve the inverse problem of the Jacobian matrix of a non-square matrix, it is unstable near the singularity and introduces further numerical inaccuracies. The elimination of singularity and improvement of stability have been studied extensively, and some of these have developed into algorithms such as KDL, TRAC-IK, and commercial optimization algorithms including SQP-double quadratic and SQP-sum of squares. Recently the rapid development of machine learning has demonstrated the huge potential to obtain the inverse model through different exploration schemes such as motor babbling (time-consuming, 1×104 times trials).

5.1 The Improved Structure Design of the Variable Cross-Section for the Trunk-like Soft Manipulator

The structure design of the improved variable section trunk-like soft manipulator is as follows.

Figure 5.2 indicates the overall view and side view of the trunk-like soft manipulator respectively. It can be seen that the overall configuration of the soft manipulator

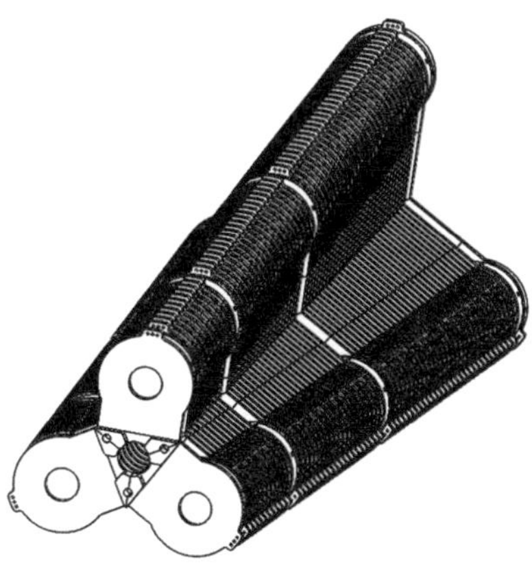

(a) Overall view

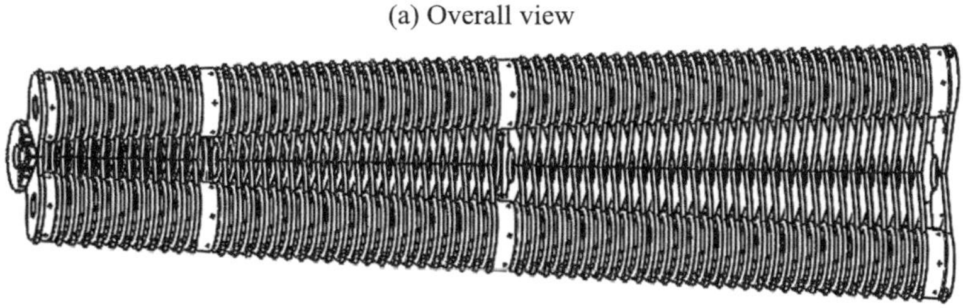

(b) Side view

Fig. 5.2 Design drawing of the trunk-like soft manipulator presents a variable cross-section, The length of the end driver is the shortest and the length of the root driver is the longest. The specific dimensions of the three aspects of the soft manipulator are shown in the Fig. 5.2.

As can be seen from Fig. 5.3, the length of the first aspect is 146.5 mm, the diameter of the widest part of the bellows cross section is 53.6 mm, and the diameter of the narrowest part is 24 mm. The distance between the center of the upper end bellows and the center of the driver is 47.9 mm, and the distance between the center of the lower end bellows and the center of the driver is 54 mm. The angle between the bellows normal and the driver normal is 3°. For the second aspect, the distance between the center of the upper end bellows and the center of the driver is 55.2 mm, and the distance between the center of the lower end bellows and the center of the driver is 67 mm. The other dimensions are the same as those of the first aspect. For the third aspect, the distance between the center of the upper end bellows and the center of the driver is 68.3 mm, and the length between the center of the lower end bellows and the center of the drive is 86 mm. The other dimensions are the same as those of the first aspect.

For a general pneumatic actuator, the end position of the actuator can be regarded as a function of the inputting pressures. Here a polynomial fitting with Taylor series was established to characterize the relationship between the inputting pressures and the outputting position/orientation as:

$$\begin{bmatrix} \boldsymbol{r} \\ \boldsymbol{\theta} \end{bmatrix} = \begin{bmatrix} \boldsymbol{r}_0 \\ \boldsymbol{\theta}_0 \end{bmatrix} + \begin{bmatrix} \sum_{i=1}^{N} \frac{\partial \boldsymbol{r}}{\partial p_i} \Delta p_i \\ \sum_{i=1}^{N} \frac{\partial \boldsymbol{\theta}}{\partial p_i} \Delta p_i \end{bmatrix} + \begin{bmatrix} \sum_{i=1}^{N} \sum_{j=i}^{N} \left(\frac{1}{2} \frac{\partial^2 \boldsymbol{r}}{\partial p_i \partial p_j} \Delta p_i \Delta p_j \right) \\ \sum_{i=1}^{N} \sum_{j=i}^{N} \left(\frac{1}{2} \frac{\partial^2 \boldsymbol{\theta}}{\partial p_i \partial p_j} \Delta p_i \Delta p_j \right) \end{bmatrix} + \cdots \tag{5.1}$$

in which the end position and orientation of the actuator is denoted as $\boldsymbol{r}$ and $\boldsymbol{\theta}$, the inputting pressures of the pneumatic actuator with multi-chambers are denoted as $\boldsymbol{p} = (p_1 \quad p_2 \quad \ldots \quad p_N)^{\mathrm{T}}$. Correspondingly, the end position and orientation of the actuator is denoted as $\boldsymbol{r}_0$ and $\boldsymbol{\theta}_0$ with the initial atmospheric pressure $\boldsymbol{p}_0 = (p_0 \quad p_0 \quad \ldots \quad p_0)^{\mathrm{T}}$, and the relative pressure can be expressed as $\Delta \boldsymbol{p} = (\Delta p_1 \quad \Delta p_2 \quad \ldots \quad \Delta p_N)^{\mathrm{T}} = (p_1 - p_0 \quad p_2 - p_0 \quad \ldots \quad p_N - p_0)^{\mathrm{T}}$.

In practice, the approximate order of the polynomial fitting can be selected according to the accuracy requirements, and the partial coefficient is obtained by numerical fitting according to the simulation or experimental data. Taking the first-order Taylor truncation of the position as an example, the linear approximation can be expressed as $\boldsymbol{r}(\boldsymbol{p}) \approx \boldsymbol{r}_0 + \partial \boldsymbol{r}/\partial \boldsymbol{p} \Delta \boldsymbol{p}$. In this sense, the modeling of the forward kinematics is actually to determine the coefficient matrix $\partial \boldsymbol{r}/\partial \boldsymbol{p}$. Figure 5.4a and b demonstrate the polynomial fitting of the position and orientation respectively. According to the deformation pattern of the actuator, different orders of the polynomial fitting can be established.

Figures in Fig. 5.4c correspond to elongation, bending, torsion and shortening respectively. In the following, we will select three typical motion modes of air-driven soft structure, i.e., expansion, bending and torsion deformation, to verify the

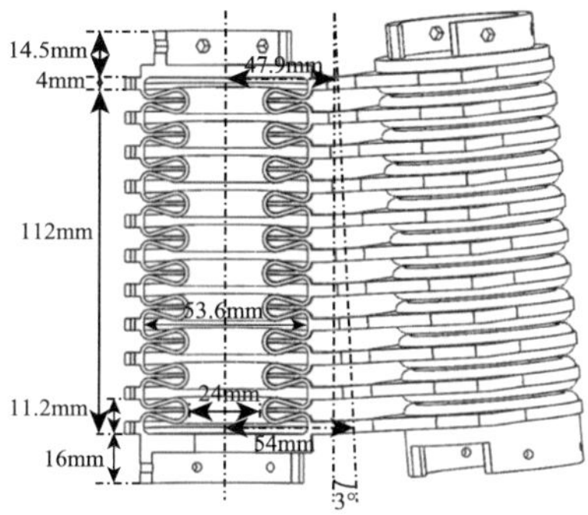

(a) The first aspect

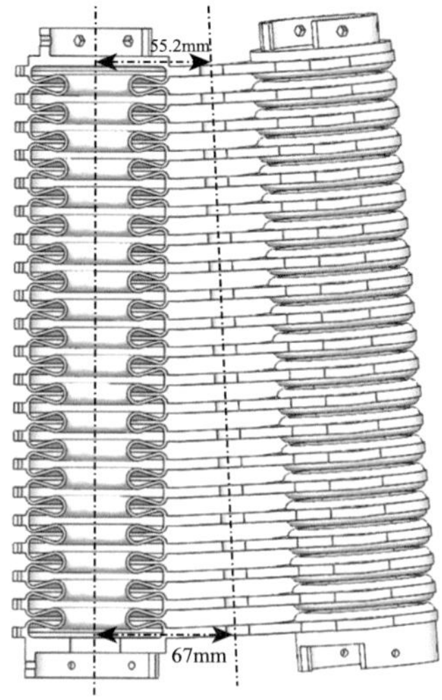

(b) The second aspect

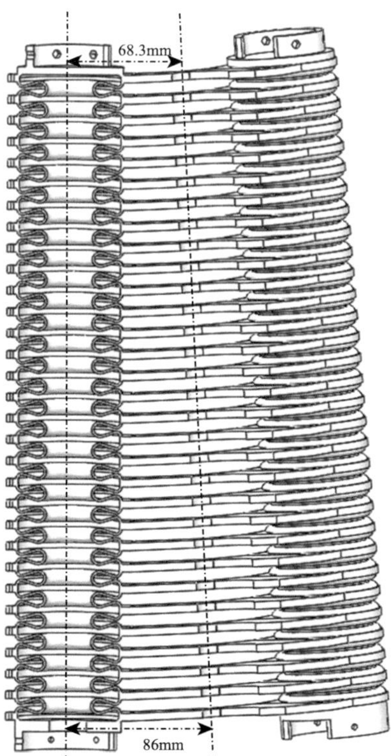

(c) The third aspect

Fig. 5.3 Design dimensions of the trunk-like soft manipulator

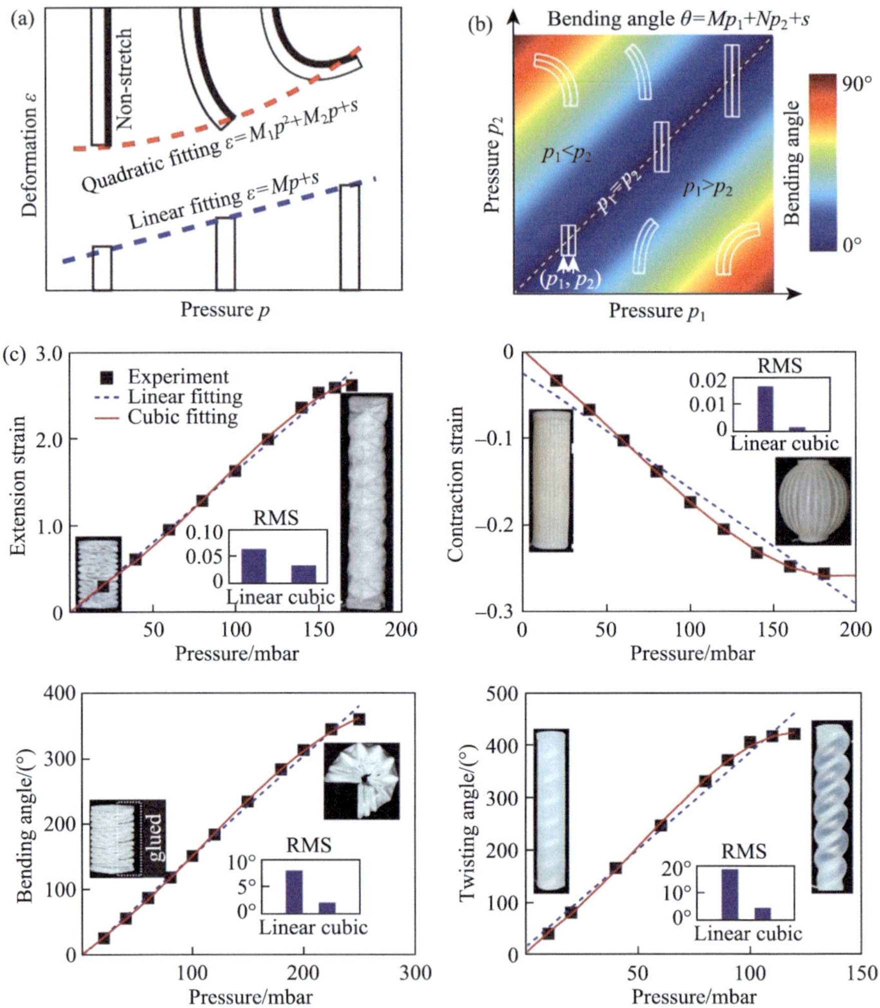

Fig. 5.4 Linear fitting and third-order fitting of single pipe manipulator in different states

universality of the forward kinematics fitting model proposed in this section, and finally apply the forward fitting model to a complex series-parallel soft manipulator driven by air.

For a given set of air pressures and the end position of a pneumatic actuator, linear fitting and quadratic fitting were used to establish the relationship between the inputting pressures and the outputting positions.

It can be seen from Fig. 5.4c that the three basic motion forms of pneumatic soft manipulator, such as stretching, bending and twisting, can be uniformly described by the series expansion method proposed in this chapter. Since the movement of arbitrary soft manipulator driven by air pressure can be regarded as the combination of basic motion forms such as stretching, bending and twisting, the series expansion method in this chapter can be used to describe the motion behavior of the other soft manipulators driven by air pressure to a certain degree.

Figure 5.4 also indicate that in the range of low air pressure, the motion of the pneumatic actuator can be well described by taking the series expansion to the first order, that is, using a linear model, which will greatly simplify the mathematical expression of the kinematic model of the soft manipulator [1, 2]. It is very suitable for the closed-loop control process of the precise movement of the manipulator.

We collected a large amount of data, including the experimental data of an air-driven flexible arm structure published by Ramses V. Martinez, Carina R. Fish and others in 2012, and the data of a three-way nine-segment air-driven flexible arm measured by our experiments, as indicated in Fig. 5.5. The fitting results were obtained by the least square method.

5.2 Path Planning and Pressure Control of the Soft Manipulator

5.2.1 *Simplified Model Based on Experimental Data*

The motion model of the soft manipulator based on the constant curvature assumption [3] is prone to be not able to obtain feasible solution because of singularity of equation when solving inverse kinematics problem [4, 5]. In the following, a simplified mechanical model of the soft manipulator is established through the method of fitting test and measurement data, so as to provide a feasible solution for the inverse problem in the process of path planning and control. The position data of the mark points of the soft manipulator can be obtained by the electronic total Station or VICON motion measurement system [6]. Here, we take the electronic total station as an example to illustrate the method of establishing the simplified mechanical model of the soft manipulator.

As shown in Fig. 5.6, $O\text{-}ENZ$ denotes the electronic total station coordinate and $o\text{-}xyz$ denotes the coordinate system of the soft manipulator. In the electronic total station coordinate system, E points to the east direction, N points to the north direction, Z points to the direction of altitude; In the coordinate system of the soft manipulator, the origin of the coordinate is located in the center of the bottom frame, the x direction is horizontal to the right, y direction points to the inside direction, z points to the height direction.

The transformation relationship between the two coordinate systems is

$$\begin{bmatrix} x \\ y \\ z \end{bmatrix} = \boldsymbol{A} \begin{bmatrix} E \\ N \\ Z \end{bmatrix} + \begin{bmatrix} E_0 \\ N_0 \\ Z_0 \end{bmatrix} \tag{5.2}$$

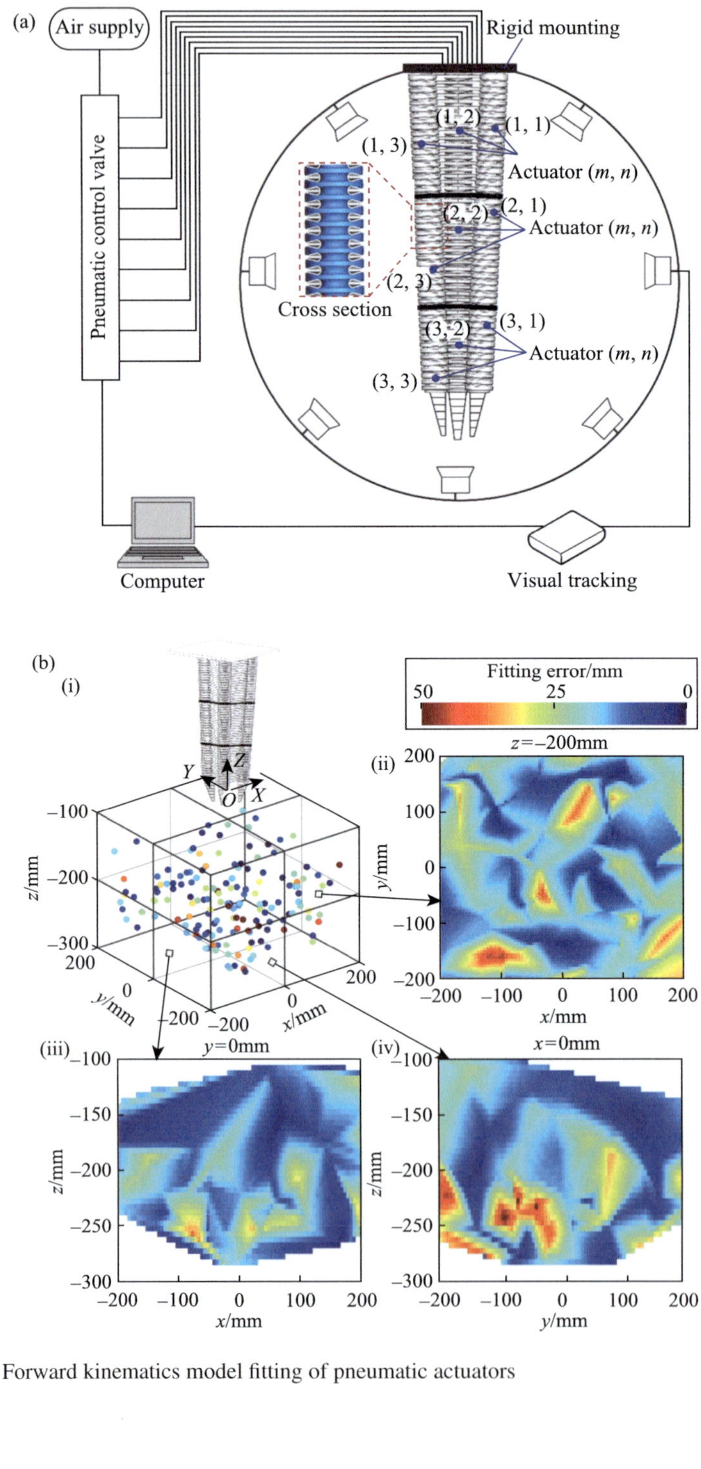

Fig. 5.5 Forward kinematics model fitting of pneumatic actuators

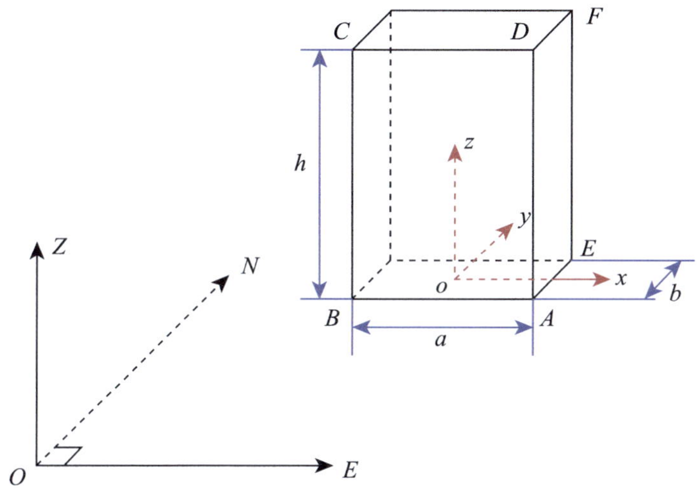

Fig. 5.6 Total station coordinate system and soft manipulator coordinate system. O-ENZ and o-xyz are both right-handed rectangular coordinate systems

where, $\boldsymbol{A}$ is a 3×3 coordinate transformation matrix. We know that $A\sim F$ in the o-xyz coordinate system are:

$$
\begin{aligned}
[x_A \quad y_A \quad z_A]^{\mathrm{T}} &= \left[\frac{a}{2} \quad -\frac{b}{2} \quad 0\right]^{\mathrm{T}} \\
[x_B \quad y_B \quad z_B]^{\mathrm{T}} &= \left[-\frac{a}{2} \quad -\frac{b}{2} \quad 0\right]^{\mathrm{T}} \\
[x_C \quad y_C \quad z_C]^{\mathrm{T}} &= \left[-\frac{a}{2} \quad -\frac{b}{2} \quad h\right]^{\mathrm{T}} \\
[x_D \quad y_D \quad z_D]^{\mathrm{T}} &= \left[\frac{a}{2} \quad -\frac{b}{2} \quad h\right]^{\mathrm{T}} \\
[x_E \quad y_E \quad z_E]^{\mathrm{T}} &= \left[\frac{a}{2} \quad \frac{b}{2} \quad 0\right]^{\mathrm{T}} \\
[x_F \quad y_F \quad z_F]^{\mathrm{T}} &= \left[\frac{a}{2} \quad \frac{b}{2} \quad h\right]^{\mathrm{T}}
\end{aligned} \tag{5.3}
$$

$A\sim F$ in the O-ENZ coordinate system can also be obtained by measurement. According to the coordinate transformation relationship between the two coordinate systems, it can be obtained that:

$$
\begin{bmatrix} x_A \\ y_A \\ z_A \end{bmatrix} = \boldsymbol{A} \begin{bmatrix} E_A \\ N_A \\ Z_A \end{bmatrix} + \begin{bmatrix} E_0 \\ N_0 \\ Z_0 \end{bmatrix} \tag{5.4}
$$

$$\begin{bmatrix} x_B \\ y_B \\ z_B \end{bmatrix} = \boldsymbol{A} \begin{bmatrix} E_B \\ N_B \\ Z_B \end{bmatrix} + \begin{bmatrix} E_0 \\ N_0 \\ Z_0 \end{bmatrix} \tag{5.5}$$

$$\begin{bmatrix} x_C \\ y_C \\ z_C \end{bmatrix} = \boldsymbol{A} \begin{bmatrix} E_C \\ N_C \\ Z_C \end{bmatrix} + \begin{bmatrix} E_0 \\ N_0 \\ Z_0 \end{bmatrix} \tag{5.6}$$

$$\begin{bmatrix} x_D \\ y_D \\ z_D \end{bmatrix} = \boldsymbol{A} \begin{bmatrix} E_D \\ N_D \\ Z_D \end{bmatrix} + \begin{bmatrix} E_0 \\ N_0 \\ Z_0 \end{bmatrix} \tag{5.7}$$

$$\begin{bmatrix} x_E \\ y_E \\ z_E \end{bmatrix} = \boldsymbol{A} \begin{bmatrix} E_E \\ N_E \\ Z_E \end{bmatrix} + \begin{bmatrix} E_0 \\ N_0 \\ Z_0 \end{bmatrix} \tag{5.8}$$

$$\begin{bmatrix} x_F \\ y_F \\ z_F \end{bmatrix} = \boldsymbol{A} \begin{bmatrix} E_F \\ N_F \\ Z_F \end{bmatrix} + \begin{bmatrix} E_0 \\ N_0 \\ Z_0 \end{bmatrix} \tag{5.9}$$

The position of $B \sim F$ minus that of point A, and we can obtain that:

$$\begin{bmatrix} x_B - x_A \\ y_B - y_A \\ z_B - z_A \end{bmatrix} = \boldsymbol{A} \begin{bmatrix} E_B - E_A \\ N_B - N_A \\ Z_B - Z_A \end{bmatrix} \tag{5.10}$$

$$\begin{bmatrix} x_C - x_A \\ y_C - y_A \\ z_C - z_A \end{bmatrix} = \boldsymbol{A} \begin{bmatrix} E_C - E_A \\ N_C - N_A \\ Z_C - Z_A \end{bmatrix} \tag{5.11}$$

$$\begin{bmatrix} x_D - x_A \\ y_D - y_A \\ z_D - z_A \end{bmatrix} = \boldsymbol{A} \begin{bmatrix} E_D - E_A \\ N_D - N_A \\ Z_D - Z_A \end{bmatrix} \tag{5.12}$$

$$\begin{bmatrix} x_E - x_A \\ y_E - y_A \\ z_E - z_A \end{bmatrix} = \boldsymbol{A} \begin{bmatrix} E_E - E_A \\ N_E - N_A \\ Z_E - Z_A \end{bmatrix} \tag{5.13}$$

$$\begin{bmatrix} x_F - x_A \\ y_F - y_A \\ z_F - z_A \end{bmatrix} = \boldsymbol{A} \begin{bmatrix} E_F - E_A \\ N_F - N_A \\ Z_F - Z_A \end{bmatrix} \tag{5.14}$$

By considering the Eqs. (5.10)–(5.14), we can obtain:

$$\begin{bmatrix} x_B - x_A & x_C - x_A & x_D - x_A & x_E - x_A & x_F - x_A \\ y_B - y_A & y_C - y_A & y_D - y_A & y_E - y_A & y_F - y_A \\ z_B - z_A & z_C - z_A & z_D - z_A & z_E - z_A & z_F - z_A \end{bmatrix} = \boldsymbol{A} \begin{bmatrix} E_B - E_A & E_C - E_A & E_D - E_A & E_E - E_A & E_F - E_A \\ N_B - N_A & N_C - N_A & N_D - N_A & N_E - N_A & N_F - N_A \\ Z_B - Z_A & Z_C - Z_A & Z_D - Z_A & Z_E - Z_A & Z_F - Z_A \end{bmatrix} \tag{5.15}$$

We define:

$$\boldsymbol{Q} = \begin{bmatrix} x_B - x_A & x_C - x_A & x_D - x_A & x_E - x_A & x_F - x_A \\ y_B - y_A & y_C - y_A & y_D - y_A & y_E - y_A & y_F - y_A \\ z_B - z_A & z_C - z_A & z_D - z_A & z_E - z_A & z_F - z_A \end{bmatrix} \tag{5.16}$$

and

$$\boldsymbol{W} = \begin{bmatrix} E_B - E_A & E_C - E_A & E_D - E_A & E_E - E_A & E_F - E_A \\ N_B - N_A & N_C - N_A & N_D - N_A & N_E - N_A & N_F - N_A \\ Z_B - Z_A & Z_C - Z_A & Z_D - Z_A & Z_E - Z_A & Z_F - Z_A \end{bmatrix} \tag{5.17}$$

We can obtain:

$$\boldsymbol{A} = \left(\boldsymbol{Q}\boldsymbol{W}^{\mathrm{T}}\right)\left(\boldsymbol{W}\boldsymbol{W}^{\mathrm{T}}\right)^{-1} \tag{5.18}$$

Then $(E_0 \quad N_0 \quad Z_0)^{\mathrm{T}}$ can be calculated by bringing coordinate transformation matrix $\boldsymbol{A}$ into Eq. (5.10). So far, the coordinate transformation relationship between the electronic total station coordinate system O-ENZ and the soft manipulator coordinate system o-xyz has been obtained.

Here, $a = 0.660$ m, $b = 0.660$ m, $c = 0.151$ m. Through the calculation above, the final obtained coordinate transformation matrix $\boldsymbol{A}$ and position offset vector $(E_0 \quad N_0 \quad Z_0)^{\mathrm{T}}$ can be described as follows:

$$\boldsymbol{A} = \begin{bmatrix} 0.998 & -0.207 & 0 \\ 0.189 & 1.014 & 0 \\ 0 & 0 & 1 \end{bmatrix} \tag{5.19}$$

and

$$\begin{bmatrix} E_0 \\ N_0 \\ Z_0 \end{bmatrix} = \begin{bmatrix} 1.190 \\ -2.238 \\ -0.003 \end{bmatrix} \tag{5.20}$$

Here, it is assumed that the pressure input and the output of the end position of the soft manipulator can be approximately considered to be linear, that is:

$$\boldsymbol{r} = \boldsymbol{Q}\boldsymbol{p} + \boldsymbol{r}_0 \tag{5.21}$$

where, matrix $\boldsymbol{Q}$ is a matrix of 3×9, and $r_0$is a constant vector of 3×1, i.e.

$$\boldsymbol{Q} = \begin{bmatrix} q_{11} & q_{12} & q_{13} & q_{14} & q_{15} & q_{16} & q_{17} & q_{18} & q_{19} \\ q_{21} & q_{22} & q_{23} & q_{24} & q_{25} & q_{26} & q_{27} & q_{28} & q_{29} \\ q_{31} & q_{32} & q_{33} & q_{34} & q_{35} & q_{36} & q_{37} & q_{38} & q_{39} \end{bmatrix} \tag{5.22}$$

$$\boldsymbol{r}_0 = [x_0 \quad y_0 \quad z_0]^{\mathrm{T}} \tag{5.23}$$

The input pressure of the soft manipulator $\boldsymbol{p}$ can be expressed as:

$$\boldsymbol{p} = \left[\, p_1 \; p_2 \; p_3 \; p_4 \; p_5 \; p_6 \; p_7 \; p_8 \; p_9 \,\right]^{\mathrm{T}} \tag{5.24}$$

The end position of the manipulator $\boldsymbol{r}$ can be expressed as:

$$\boldsymbol{r} = \left[\, x \; y \; z \,\right]^{\mathrm{T}} \tag{5.25}$$

According to the design scheme, it can be obtained through the test and measurement results:

$$\boldsymbol{Q} = \begin{bmatrix} 0.0754 & 0.2217 & -0.3125 & 0.0871 & 0.2359 & -0.2546 & 0.0250 & 0.0649 & -0.0758 \\ -0.3011 & 0.2209 & 0.1189 & -0.3213 & 0.2032 & 0.1061 & -0.0847 & 0.0699 & 0.0189 \\ -0.0057 & -0.0291 & -0.0925 & 0.0173 & -0.0142 & -0.1261 & -0.0358 & -0.0358 & -0.0364 \end{bmatrix} \tag{5.26}$$

$$\boldsymbol{r}_0 = \left[\, -0.6991 \; -3.8822 \; 0.5359 \,\right]^{\mathrm{T}} \tag{5.27}$$

where, the unit of the manipulator input pressure $\boldsymbol{p}$ is bar and the unit of the position vector is $\boldsymbol{m}$.

5.2.2 *Path Planning and Control Based on Optimization Algorithm*

5.2.2.1 Inverse Kinematics and Path Planning of the Soft Manipulator

The problem of path planning [7] and pressure control of a soft manipulator can be described as: the end of the soft manipulator moves along the given path under the control of the air pressure. The path planning problem can be decomposed into two steps: ① calculating the pressure value of the nine air chambers of the soft manipulator according to the position of a specified single point, which is essentially the inverse problem of the soft manipulator kinematics model; ② on the premise of smooth motion and efficient vibration suppression of the manipulator, the specified motion path is discretized into several position sequences, and a set of pressure sequences of the soft manipulator are calculated according to the inverse problem of kinematics. Finally, the nine air chambers of the soft manipulator are charged according to the time series of the above pressure, so as to move the end of the soft manipulator along the specified path.

Firstly, consider the first step to calculate the pressure of the nine air chambers of the soft manipulator based on the position of a specified single point. According to the mechanical model of the soft manipulator, the input is the pressure of nine air chambers, and the output is the position vector of the end of the manipulator (there are 3 in total). Given three components of the end position of the manipulator, nine output variables are needed to be solved, which is a multi-solution problem. Here, an optimization algorithm is used to present a feasible solution. The variables of the optimization problem is the pressure of nine chambers. The objective function is to minimize the distance between the end of the manipulator and the target position and the nine pressures are required not to exceed a certain threshold to ensure the safety of the structure. The above optimization problem can be mathematically described as follows.

$$\begin{aligned} &\text{Find} && \boldsymbol{p}^{\mathrm{m}} \\ &\text{minimize} && |\boldsymbol{r}(\boldsymbol{p}^{\mathrm{m}}) - \boldsymbol{r}_t|^2 \\ &\text{subject to} && 0 \leqslant p_i^{\mathrm{m}} \leqslant p_{\max}, \quad i = 1, 2, \ldots, 9 \end{aligned} \tag{5.28}$$

where, $\boldsymbol{r}_t$ is the target position; $\boldsymbol{r}(\boldsymbol{p}^{\mathrm{m}})$ denotes the position vector of the manipulator under the pressure $\boldsymbol{p}^{\mathrm{m}}$, which can be obtained from the mechanical model of the soft manipulator. The optimization process can be regarded as the inverse problem of the mechanical model of the soft manipulator, i.e., calculating the corresponding pressure with the deformation of the soft manipulator.

Secondly, considering the second step, according to the specified path, a set of pressure series of the soft manipulator is calculated based on the above single point optimization. As shown in Fig. 5.7, the given motion path is discretized into a set of position sequences, i.e.

$$\boldsymbol{r}_k, \ k = 1, 2, \ldots, n \tag{5.29}$$

For a given target position vector, a feasible solution $\boldsymbol{p}^{\mathrm{m}}_{(k,\mathrm{optimized})}$ can be obtained through the mechanical inverse problem optimization algorithm in the first step but this solution is generally not unique. In order to ensure that the pressure changes between the two adjacent discrete positions will not be too large during the continuous movement of the soft manipulator, the initial optimization value of the current step is selected as the optimization result of the previous step, that is

$$\boldsymbol{p}^{\mathrm{m}}_{(k,\mathrm{optimized})} - \boldsymbol{p}^{\mathrm{m}}_{(k-1,\mathrm{optimized})} \tag{5.30}$$

The specific optimization process is shown in Fig. 5.7.

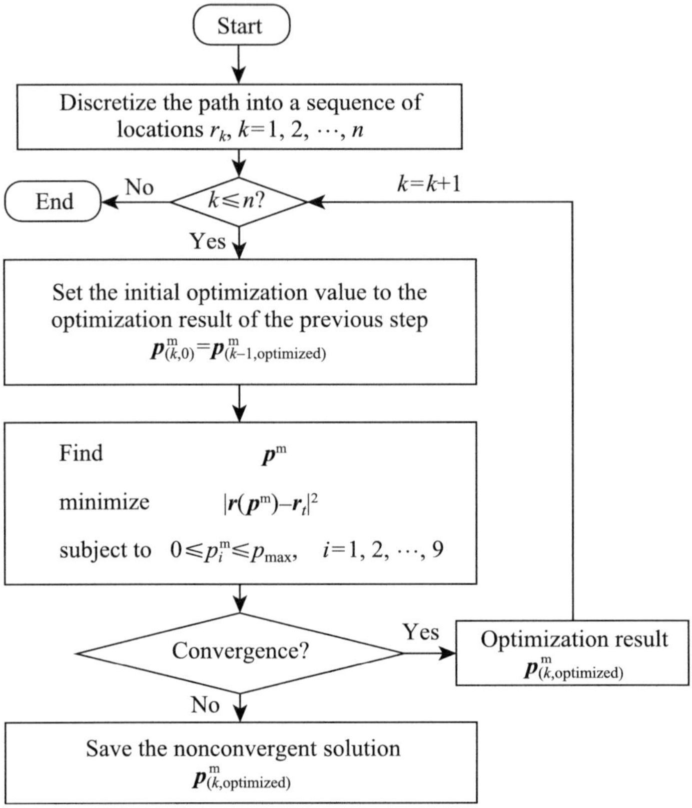

Fig. 5.7 Path planning of inflatable soft manipulator

5.2.2.2 Soft Manipulator Error Direction Control

When the end of the soft manipulator moves near the target position, the position of the end of the soft manipulator needs to be precisely adjusted based on the error direction control, so as to reduce the deviation between the end of the soft manipulator and the target position as much as possible, and realize the accurate adjustment of the end position of the soft manipulator. In practice, VICON motion capture system is used to collect the real-time position of the end of the soft manipulator, and the data is transmitted to the workstation through the data acquisition processor. The workstation obtains the position deviation of the end of the soft manipulator according to the target position and the real position of the end of the soft manipulator at present, which can be expressed as follows:

$$\boldsymbol{\Delta} = \boldsymbol{r}(\boldsymbol{p}) - \boldsymbol{r} \tag{5.31}$$

The end error accuracy of the soft manipulator is defined as ε. If $\Delta > \varepsilon$, the current end error of the soft manipulator is too large to meet the requirement of the control accuracy, and it needs to be adjusted accurately. If $\Delta \leqslant \varepsilon$, the end error of the current soft manipulator is already within the required range of control accuracy, and precise adjustment is no longer necessary. The specific implementation scheme of precise adjustment is as follows: the position deviation of the end of the soft manipulator is introduced into the optimization target, so that the position of the end of the soft manipulator is constantly approaching the target position, where the position deviation correction coefficient is β. Therefore, the optimization of the closed-loop control can be described as:

$$\begin{aligned} &\text{Find} \quad \boldsymbol{p} \\ &\text{minimize} \ |\boldsymbol{r}(\boldsymbol{p}) - \boldsymbol{r} + \beta \boldsymbol{\Delta}|^2 \\ &\text{subject to} \ 0 \leqslant p \leqslant p_{\max}, \quad i = 1, 2, \ldots, N \end{aligned} \tag{5.32}$$

The optimization objective function introduces the deviation between the current position and the target position. In the specific execution, the workstation needs to detect whether the real-time deviation has met the requirements before each execution of the optimization program. If it does not meet the requirements, the optimization program will be executed and the optimization result is used as the air pressure input. Then iterate until the deviation meets the requirements.

It should be pointed out that the error direction control method does not require the position of the end of the soft manipulator to coincide with the target position, but regards the position deviation between the end of the soft manipulator and the target as the optimization objective. The advantage of this is to ensure that a convergent solution can be obtained for each feedback control in the closed-loop control process. Considering that the target position may be outside the motion envelope of the end of the soft manipulator in practice, this control strategy can minimize the deviation from the target position while ensuring the convergence of the motion, and improve the robustness of the closed-loop tracking system.

In order to verify the control effect of the closed loop control algorithm above, the closed-loop control implementation scheme of the soft manipulator driven by air chambers is given below, as indicated in Figs. 5.8 and 5.9. The position accuracy error of closed-loop control is set to 1 cm. It can be seen that after the rapid positioning in the first step, the deviation between the real position of the end of the soft manipulator and the target position is about 200 mm; after 2–5 steps precise correction, the deviation between the real position of the end of the soft manipulator and the target position is reduced from 200 to 3 mm, which meets the given precision requirements. Summarizing the whole closed-loop control process, a total of five iterative steps are required, i.e., one step fast position process and four steps accurate correction. The inverse kinematics model can approach to the target position in a short time, but the position error is relatively large. On this basis, the error direction control achieves the manipulator's accurate arriving at the target position.

For solving the inverse problem of soft pneumatic actuators, we present a novel formulation and calculating the inverse kinematics utilizing optimization. Geomet-

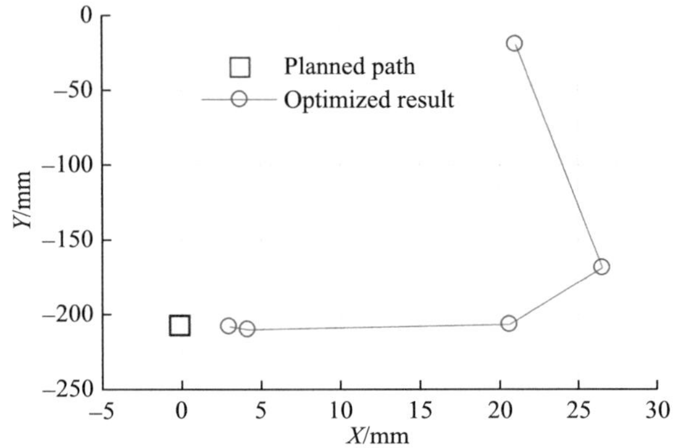

Fig. 5.8 Error direction control of soft inflatable manipulator

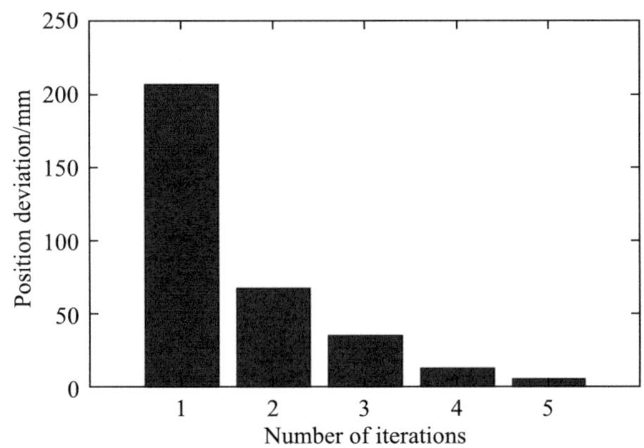

Fig. 5.9 Convergence of error direction control

rically, the aim is to find a vector of inputting pressure which drives the effector of the actuator to reach a target location. Besides, it should be taken into account that structural strength of pneumatic actuators and other restrictive boundaries such as avoidance of environmental obstacles, which result in additional constraints of design variables. The inverse problem of pneumatic actuators was formulated as a constrained optimization as follows

$$\begin{aligned} &\text{Find} \quad \boldsymbol{p} \\ &\text{minimize} \ |\boldsymbol{f}(\boldsymbol{p}) - \boldsymbol{r}^*|^2 \\ &\text{subject to} \ 0 \leqslant p_i \leqslant p_{\max}, \quad i = 1, 2, \ldots, N \\ &\qquad\qquad \boldsymbol{f}_{\text{other}}(\boldsymbol{p}) \leqslant \boldsymbol{r}_{\text{c}} \end{aligned} \tag{5.33}$$

where, j indicates the section index along the direction from the fixed base to the operating end ranging from 1 to N; i indicates the chamber index of the pneumatic actuator ranging from 1 to 3 N considering that each section contains three chambers in parallel. As a result, the position of the end of the j-th section can be expressed as $\boldsymbol{f}_j(\boldsymbol{p})$, and the effector of the actuator can be expressed as $\boldsymbol{f}_N(\boldsymbol{p})$ when $j = N$.

The objective function is to minimize the square norm of positional deviation between the operating end $\boldsymbol{f}_N(\boldsymbol{p})$ and the target location $\boldsymbol{r}_{\text{targ}}$, in which the forward kinematic model $\boldsymbol{r} = \boldsymbol{f}(\boldsymbol{p})$ is utilized. It is worth mentioning that here the position of the effector is not required to coincide with the target, but to minimize the position deviation between the effector and the target, which ensures a converged solution no matter where the target locates. More specifically, when the target is given within the reachable envelope of the effector, the inverse solution ensures that the effector coincide with the target. However, when the target is given out of the reachable envelope of the effector, the effector is closest to instead of coincide with the target.

For the structural constraint, a simple restriction $0 \leqslant p_i \leqslant p_{\max}$ is adopted in Eq. (5.33) in which the inputting pressures should be no larger than a certain threshold value $p_{\max}$, and it is worth mentioning that this can be replaced with other complicated formulations. Besides, the geometrical constraints should also be considered including the operating space and configuration of the actuator, and thus here we introduced the constraints of position of each section $\boldsymbol{C}(\boldsymbol{f}_j(\boldsymbol{p})) \leqslant \boldsymbol{0}$ for simplicity.

The optimization problem formulated by Eq. (5.33) is nonlinear with respect to the pressure vector $\boldsymbol{p}$ considering the non-unique solutions for the redundant model. The problem is solved here by using the sequential quadratic programming (SQP) algorithm [8], implemented in MATLAB 2015a.

According to the optimization algorithm of inverse kinematics, the kinematic planning and motion control of a redundant pneumatic actuator are obtained naturally. For effector positioning, the required air pressures $\boldsymbol{p}_{\text{req}}$ can be calculated through Eq. (5.33) when the target location of the actuator effector $\boldsymbol{r}_{\text{targ}}$ is given. In practice, the sequence of inputting pressures is usually smoothly interpolated from the initial $\boldsymbol{p}_{\text{init}}$ to the required $\boldsymbol{p}_{\text{req}}$ to reduce the severe impact of the airflow variation.

Furthermore, the positioning of a continuous route can be achieved by dispersing the continuous path into a series of discrete target positions $\boldsymbol{r}_k (k = 1, 2, \ldots, n)$, and the discrete spacing depends on the accuracy requirement. For each step, it is natural to obtain the air pressures through optimization, which is similar to effector positioning. Considering the multi-solution of the inverse problem, there may exist various optimal solutions for each individual step. For the series of optimizations of route planning, the optimized air pressure in the previous step is set as the initial value of the next step, which contributes to reducing the variation of the pressures between the adjacent steps, so as to reduce the structural vibration caused by the change of air pressure.

For effector positioning, only a single process of optimization is required to obtain the air pressure, which greatly simplifies the calculation of the inverse kinematics, reduces the response time, and ensures the quick positioning of the effector. For route planning, the variation of air pressure among the neighboring steps can be restrained via the correlation between the adjacent optimization, contributing to the suppression of structural vibration.

Soft robots with elastic elements usually face additional problems with nonstationary behaviors due to hysteresis and wear out effects of the mechanically exposed material. For the pneumatic actuators studied in this chapter, the nonstationary behaviors are mainly caused by structural damping and the friction between the piping system and the structure. In addition, the error of both the inverse kinematics model and the air pressure control will make it difficult to realize accurate control for the effector positioning and route planning. In order to improve the positional precision, here a closed-loop control strategy is proposed for the effector positioning and route planning, as indicated in Fig. 5.10.

For the closed-loop control, the real-time positions of the effector and target are measured and preprocessed by a VICON motion capture system, and collected by a workstation. According to the measured results, the vector of position deviation from the target to the effector of the actuator can be expressed as

$$\boldsymbol{\Delta} = \boldsymbol{f}(\boldsymbol{p}) - \boldsymbol{r} \tag{5.34}$$

The position deviation $|\boldsymbol{\Delta}|$ is introduced into the objective function as the correction term as seen in Eq. (5.34), and the effector is continuously approached to the target by the closed-loop control. In this section, a simple proportional feedback is adopted as an example to demonstrate the closed-loop control, in which β is the proportional coefficient. The closed-loop control based on the optimization can be expressed as

$$\begin{array}{ll} \text{Find} & \boldsymbol{p} \\ \text{minimize} & |\boldsymbol{f}(\boldsymbol{p}) - \boldsymbol{r}^* + \beta\boldsymbol{\Delta}|^2 \\ \text{subject to} & 0 \leqslant p_i \leqslant p_{\max}, \quad i = 1, 2, \ldots, N \\ & \boldsymbol{f}_{\text{other}}(\boldsymbol{p}) \leqslant \boldsymbol{r}_{\text{c}} \end{array} \tag{5.35}$$

Before the implementation of the above feedback adjustment, the position deviation needs to be measured according to Eq. (5.33) to check whether the current deviation $|\boldsymbol{\Delta}|$ has met the requirement ε. In detail, the feedback adjustment in Eq. (5.34) needs to be carried out when $|\boldsymbol{\Delta}| > \varepsilon$, and the interaction of the closed-loop control stops until $|\boldsymbol{\Delta}| > \varepsilon$.

Similar to the open-loop control, the tracking of a continuous route can also be achieved by the proposed closed-loop control, in which each discrete target can be located through the iteration of Eq. (5.34). For both the effector positioning and the continuous tracking, the closed-loop control significantly improves the positional accuracy.

5.3 Principle Prototype and Test System

5.3.1 Principle Prototype Design and Development

The principle prototype of the soft manipulator is composed of the following six parts: soft robot arm (9 independent gas path control), end soft gripper (2 independent gas path control), upper computer (control software), controller (Beckhoff TwinCAT), valve terminal (FESTO Motion Terminal), air compressor (maximum

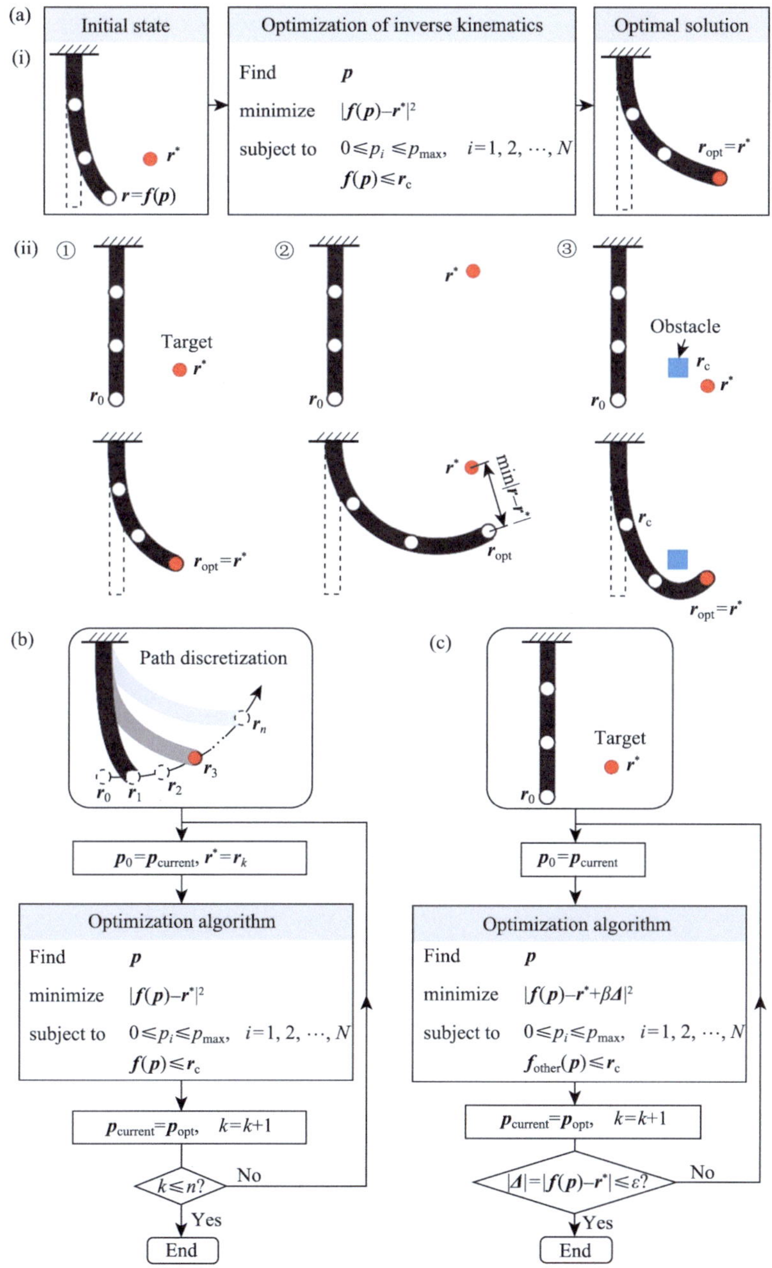

Fig. 5.10 The path planning and control process of the soft manipulator

available atmospheric pressure: 8 bar). The overall structure of the system is shown in Fig. 5.11.

In Fig. 5.11, the three soft drivers of the soft robot arm are all integrated through nylon glass fiber laser sintering, and then the three soft drivers are assembled to form the principle prototype of the soft robot arm, which is controlled by nine independent gas paths. The end soft gripper adopts the structural adaptive deformation design, which can produce adaptive deformation according to different external surface configurations of the object, so as to achieve the enveloping grasp. It is controlled by two independent air paths. The controller adopts Beckhoff TwinCAT control system, the valve terminal adopts FESTO Motion Terminal pressure control system. The controller and valve terminal communicate with each other through EtherCAT protocol. The upper computer uses Beckhoff TwinCAT programming language, which can control several fixed modes of the soft manipulator. The operator can carry out the fixed-point target grasping experiment by inputting position coordinates. The principle of upper computer control will be introduced in detail in the next section.

5.3.2 *Upper Computer Control System*

The operating interface of the upper computer control system mainly includes three parts. The left module has three typical motion modes, including spiral winding, four-leaf clover plane motion, and pendulum swinging from side to side. Each motion mode corresponds to a set of given air pressure sequences. Taking spiral motion as an example, the nine air chambers of the soft manipulator need to provide 9 air pressure changing sequences, that is:

$$\boldsymbol{p} = \begin{bmatrix} p_1(t) & p_2(t) & p_3(t) & p_4(t) & p_5(t) & p_6(t) & p_7(t) & p_8(t) & p_9(t) \end{bmatrix}^{\mathrm{T}} \tag{5.36}$$

In practice, the air pressure sequence $\boldsymbol{p}$ is a set of discrete barometric values, and the time interval between each set of barometric values is Δt. Specifically, the time interval given here is $\Delta t = 50$ ms. The intermediate module corresponds to the fixed-point capture function of the soft manipulator. According to the position coordinates of the target point, a set of pressure control sequences of the nine air champers of the soft manipulator can be obtained by the optimization algorithm, that is, the pressure sequence of the soft manipulator:

$$\boldsymbol{p}_{\mathrm{m}} = \begin{bmatrix} p_1 & p_2 & p_3 & p_4 & p_5 & p_6 & p_7 & p_8 & p_9 \end{bmatrix}^{\mathrm{T}} \tag{5.37}$$

It should be pointed out that the pressure control sequence of the motion process of the soft manipulator moving from the initial position to the target position is obtained by linear interpolation method [9], that is:

$$\boldsymbol{p}_{\mathrm{m}}(t) = \frac{n\Delta t}{T}\boldsymbol{p}_{\mathrm{m}} \tag{5.38}$$

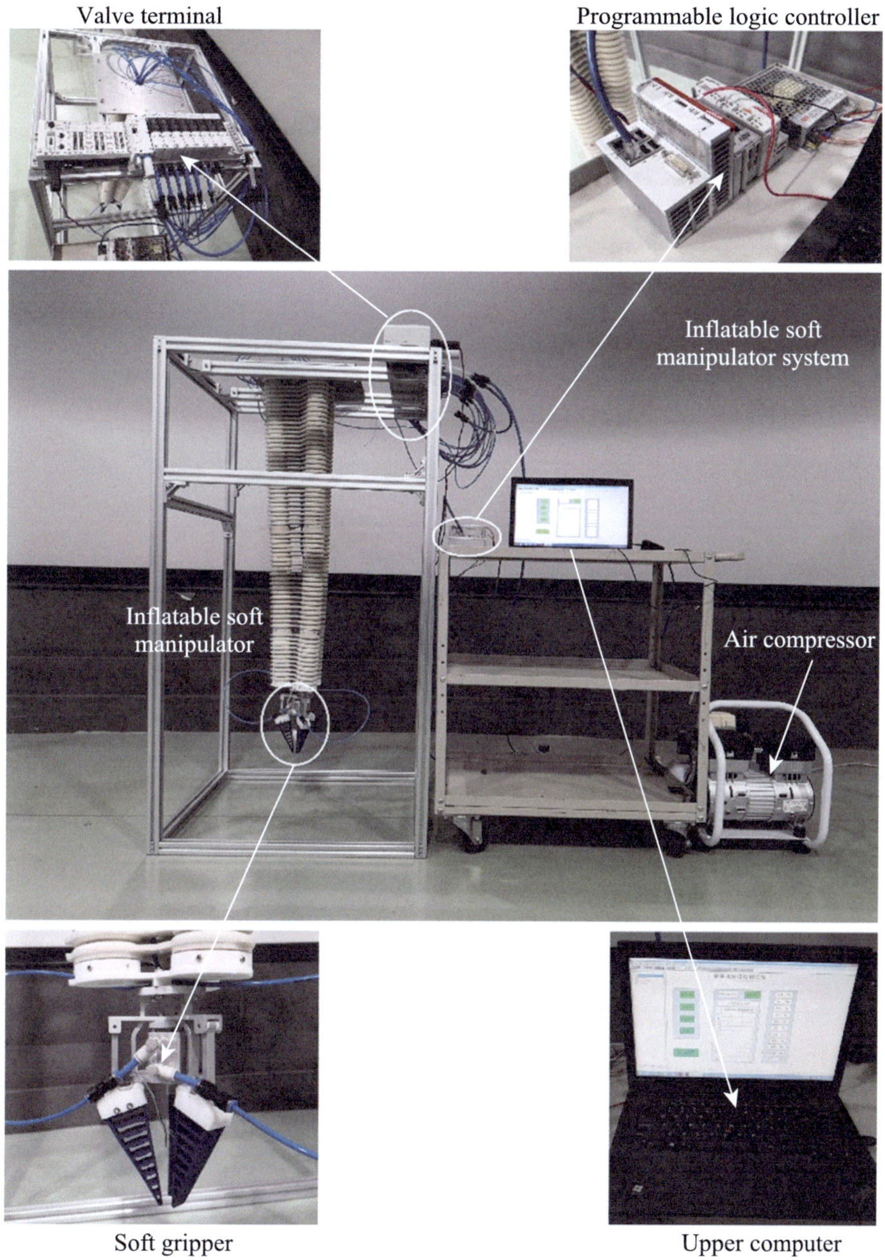

Fig. 5.11 Principle prototype and test system of the soft manipulator

where, Δt is the time interval; T is the total time of movement; n is the number of the current sequence. Under the action of the pressure sequence $\boldsymbol{p}_{\mathrm{m}}(t)$ of the soft manipulator, the end of the soft robot manipulator moves to the target position. After that, the end soft gripper starts, and the pneumatic control system provides 2 air gripper pressure control, that is

$$\boldsymbol{p}_{\mathrm{g}}(t) = \left[\, p_{10}(t)\ p_{11}(t) \,\right]^{\mathrm{T}} \tag{5.39}$$

The air pressure can realize the positive and negative movement of the cylinder, so as to drive the opening and closing operation of the soft gripper, and realize the grasping operation of the fixed point target. The right module is a real-time air pressure display module. Valves 1–11 display the air pressure values (unit: mbar) of the 9 air chambers of the current soft manipulator and the 2 air chambers of the soft gripper l.

Validations of the inverse kinematics were demonstrated in this section, in which a motion control system was built including 5 parts: a pneumatic robot arm, a grasping mechanism, a motion measurement system, a workstation, and a pressure control system with a gas pump. The introduction of each part is described as follows. 1 The soft pneumatic actuator is able to stretch and bend in the three-dimensional space to reach the target or track the continuous path. 2 The gripper locates at the end of the actuator and achieves the operation of capture/release. 3 The motion measurement system provides the workstation with the real-time position coordinates of the actuator and the target. 4 The workstation serves as the core processor to run the proposed optimization algorithm of inverse kinematics to provide the pressure input commands. For an open-loop control, the optimization algorithm of inverse kinematics is performed by the workstation to obtain the vector of the input pressures. For a closed-loop control, the feedback control based on the optimization algorithm needs to be performed in which the real-time position coordinates are provided by the motion measurement system. 5 The pressure control system receives the pressure input commands from the workstation, and adjusts the multi-channel pressure into different chambers from the gas pump.

The pneumatic robot arm adopted in this chapter is inspired by elephant trunks and fabricated by 3D printing with polyamide. The pneumatic robot arm is composed of three segments in series, and each segment contains three pneumatic bellows in parallel, resulting in $n = 9$ channels of air pressure control. Each bellow chamber can be filled with compressed air, which extends the length of the axis. When the three bellows of each segment are inflated synchronously, the symmetry along the axis of the structure assures the linear elongation. A deformation of bend can be achieved with asynchronous inflation among the bellows of each segment. Besides, the grasping mechanism is composed of three symmetrically distributed fingers, each of which can generate adaptive deformation according to the different shapes of the object to be captured. Operations including the capture and release can be driven by two input air pressures, which are also governed by the pressure control system. Therefore, there are total 11 independent pressure channels within the pressure control system as follows:

$$\begin{aligned} \boldsymbol{p}_{\mathrm{m}} &= \begin{bmatrix} p_1 \ p_2 \ p_3 \ p_4 \ p_5 \ p_6 \ p_7 \ p_8 \ p_9 \end{bmatrix}^{\mathrm{T}} \\ \boldsymbol{p}_{\mathrm{g}} &= \begin{bmatrix} p_{10} \ p_{11} \end{bmatrix}^{\mathrm{T}} \end{aligned} \tag{5.40}$$

in which $\boldsymbol{p}_{\mathrm{m}}$ indicates the 9 channels for the control of the soft robot arm; $\boldsymbol{p}_{\mathrm{g}}$ indicates the 2 channels for the control of the grasping mechanism.

Here an interactive interface of the pressure control system was designed as Fig. 5.12 shows. Three different types of movements were implanted in the left column including a helical motion, a four-leaf clover motion, and a fixed-point capture. The pressure data needed to achieve these movements were solved based on the optimization algorithm of inverse kinematics. Besides, a functional zone of "Real-time read" is listed in the middle column so that the pressure data generated by the workstation is read in real time. The right column is the real-time display of air pressure among the chambers with the unit mbar.

Based on the above forward fitting model, the optimization algorithm of inverse kinematics proposed in this chapter is applied to motion control of the pneumatic robot arm to examine its feasibility. Three examples of movement were conducted including a helical motion, a four-leaf clover motion, and a fixed-point capture, as indicated in Fig. 5.13. The first two examples accomplish an open-loop control in which the optimization algorithm of inverse kinematics runs only once without iteration if the target is given, while the last example achieves a closed-loop control in which the optimization algorithm is called and executed for several times to precisely adjust the position.

The first example is the route tracking along a helical curve. The initial position of the effector is $\boldsymbol{r}_0 = [\,x_0 \ y_0 \ z_0\,]^{\mathrm{T}} = [\,0.3070 \ 0.3120 \ 0.5418\,]^{\mathrm{T}}$m when the pneumatic actuator is not inflated, and the displacement relative to the initial position can be expressed as $\boldsymbol{r}_d = [r\cos\theta \quad r\sin\theta \quad -\theta h/(6\pi) - d]^{\mathrm{T}}$ with the radius of the helical curve $r = 0.2$m, the initial offset along the vertical direction is $d = 0.05$m, and the height variation of the helical curve is $h = 0.30$m. As a result, the analytical equation of the helical route in this example can be expressed as:

$$\boldsymbol{r} = \boldsymbol{r}_0 + \boldsymbol{r}_d = \begin{bmatrix} x_0 \\ y_0 \\ z_0 \end{bmatrix} + \begin{bmatrix} r\cos\theta \\ r\sin\theta \\ -\theta h/(6\pi) - d \end{bmatrix}, \quad \theta \in [0, 6\pi] \tag{5.41}$$

The effector will move from top to bottom along the specified spiral curve with 3 cycles in total. Each cycle will drop 0.10 m along the vertical direction, resulting in 0.30m height reduction during the whole movement.

Firstly, the continuous helix route was discretized into a series of discrete target positions. For a given target, the optimization algorithm was used to solve a set of input pressure, so that the distance between the effector and the target is minimized. In order to ensure that the air pressure at two adjacent discrete steps will not vary to a large extent during the whole movement, the initial optimization value of the current step is selected as the optimal result of the previous step. After a series of optimization, the final optimization result is shown as the hollow dot in the figure.

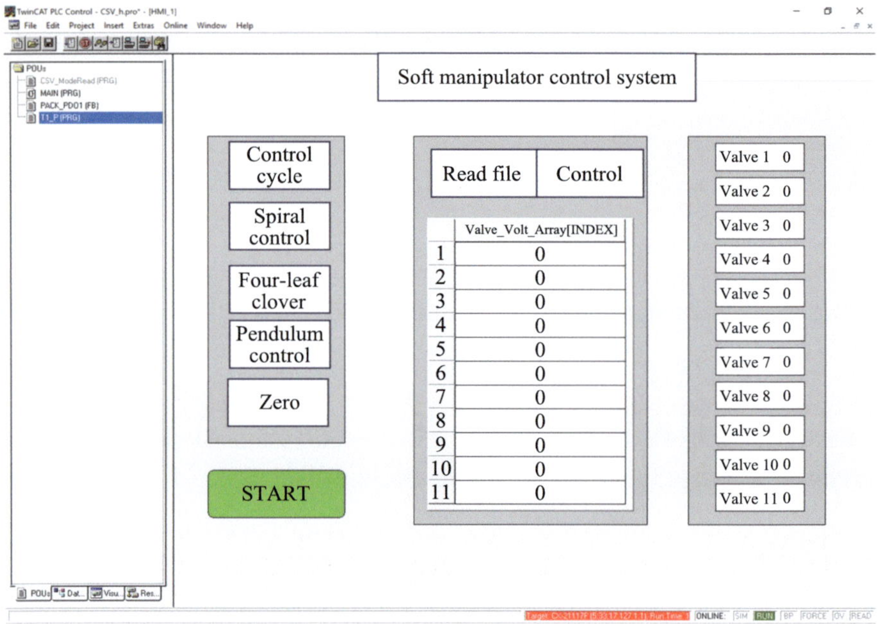

Fig. 5.12 Software manipulator control system operation interface

Considering that only the route tracking was involved in this example without capture, only the timing series of $\boldsymbol{p}_{\mathrm{m}}$ needs to be solved and executed in the soft robot arm. The positions of the effector were measured by the VICON tracking system and plotted in figure. Configurations of the soft pneumatic arm at four different moments are captured in each spiral cycle. It can be seen that the effector accomplished the circular motion from the vertical view and drops continuously along the vertical direction.

The second example is the route tracking along a four-leaf clover curve. The initial position of the effector is $\boldsymbol{r}_0 = [x_0 \quad y_0 \quad z_0]^{\mathrm{T}} = [0.3070 \quad 0.3120 \quad 0.5418]^{\mathrm{T}}\mathrm{m}$ when the pneumatic actuator is not inflated, and the four-leaf clover curve relative to the initial position can be expressed as $\boldsymbol{r}_{\mathrm{d}}$ with the maximum radial length $r = 0.50$ m and he initial offset along the vertical direction is $d = 0.05$m. As a result, the analytical equation of the four-leaf clover curve in this example can be expressed as:

$$\boldsymbol{r} = \boldsymbol{r}_0 + \boldsymbol{r}_{\mathrm{d}} = \begin{bmatrix} x_0 \\ y_0 \\ z_0 \end{bmatrix} + \begin{bmatrix} \cos(\pi/4) & \sin(\pi/4) & 0 \\ -\sin(\pi/4) & \cos(\pi/4) & 0 \\ 0 & 0 & 1 \end{bmatrix} \cdot \begin{bmatrix} r\sin(2\theta)\cos\theta \\ r\sin(2\theta)\sin\theta \\ d \end{bmatrix}, \quad \theta \in [0,\ 2\pi] \tag{5.42}$$

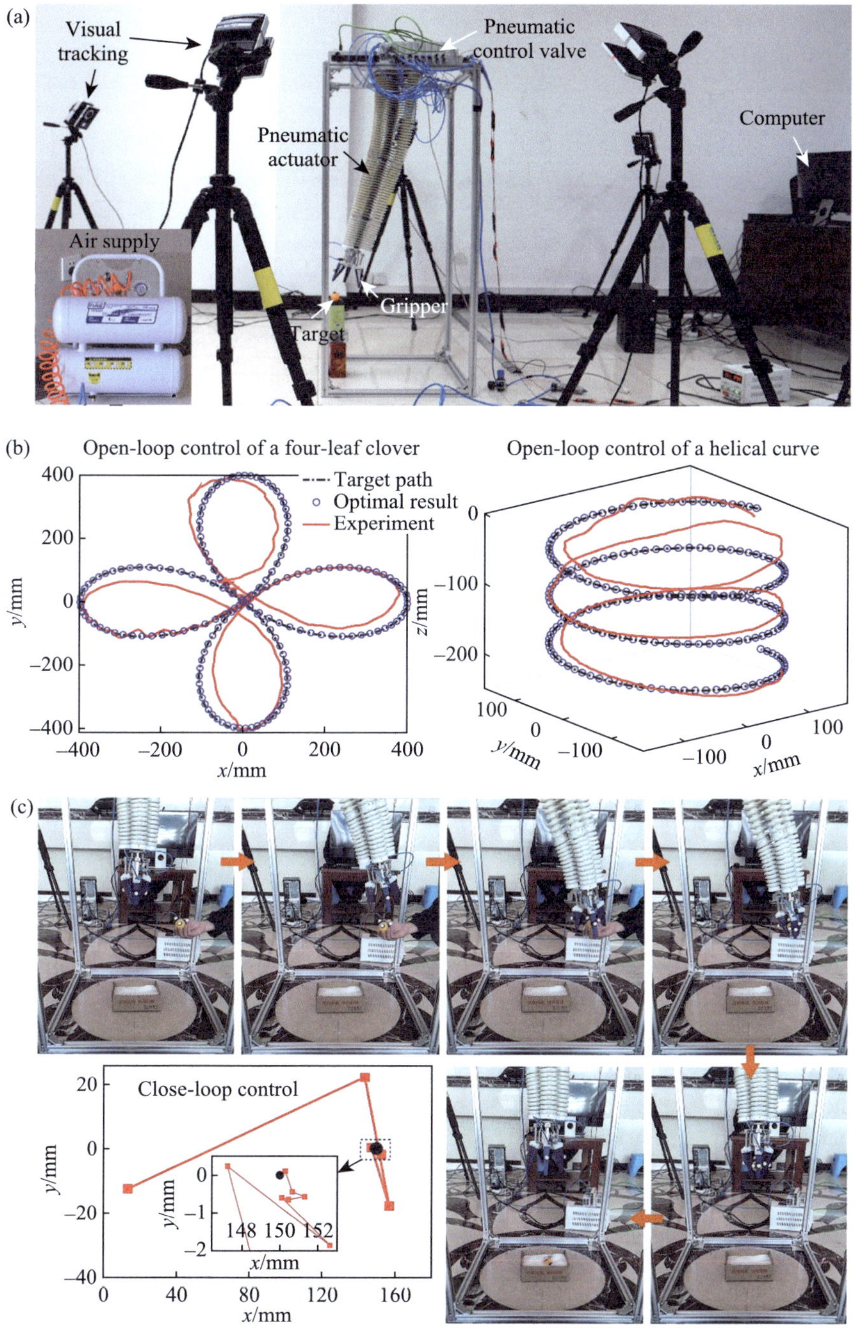

Fig. 5.13 Results of the closed-loop control

The positions of the effector were measured by the VICON tracking system and plotted in figure. Configurations of the soft pneumatic arm are captured as figure shows. It can be seen that the effector accomplished the four-leaf clover curve from the vertical view.

The third example is about a fixed-point capture, in which the effector of the soft pneumatic arm is supposed to grasp a target object (a table tennis ball) when the object is placed at a fixed position $\boldsymbol{r}_{\text{targ}}$. Given the position of the target object, the input pressures of the pneumatic arm $\boldsymbol{p}_{\text{m}}$ can be solved based on the optimization algorithm of inverse kinematics. To reduce the structural vibration caused by the huge variation of the pressure to the chambers, here a linear interpolation smoothing of the input pressures was taken with the number of interpolation points N and the time interval is $\Delta t = 20$ ms. Under the above sequence of input pressures, the effector of the pneumatic arm moved near the target object.

After the above open-loop control, a closed-loop control was used to achieve the precise adjustment of the effector. The real-time positions of both the effector and the target were measured by a VICON tracking system and collected by a workstation. According to the position deviation between the target and the effector, the workstation updated the input pressures based on the closed-loop control proposed by until the position deviation is within the accuracy. Next, two input pressures of the gripper $\boldsymbol{p}_{\text{g}}$ were controlled to achieve the capture operation of the target object while the $\boldsymbol{p}_{\text{m}}$ remained unchangeable. Once the capture was finished, the input pressures of the pneumatic arm $\boldsymbol{p}_{\text{m}}$ decreased to zero with a linear interpolation smoothing so that the effector recovered its initial position $\boldsymbol{r}_0$.

Results of the closed-loop control were plotted and compared in figures when two kinds of position accuracy were set as 10 and 1 mm. For the first case with 10 mm accuracy, the position deviation between the effector and the target reduced from 200 to 70 mm when the optimization algorithm was conducted as the first step, followed with the deviation reduction from 70 to 3 mm when the closed-loop control was conducted as steps 2–4. To sum up, the open-loop control achieves a quick but rough movement, while the closed-loop control ensures an accurate adjustment with several iterations. Compared with the former example with 10 mm precision, the example with 1 mm precision required a larger number of iterations to complete the closed-loop control. As can be seen from figure, the position deviation between the effector and the target reduced from 150 to 75 mm when the open-loop control was conducted, followed by the last six steps of the closed-loop control (steps 2–7) with the deviation reduction from 75 to 0.6 mm.

5.4 Soft Manipulator Control Test

The following ground experiments show several typical motion modes of the soft manipulator, including spiral winding, four-leaf clover movement, and pendulum swinging from side to side. In order to eliminate the influence of ground gravity, the oscillating plane experiment is carried out on the horizontal plane. Further, through

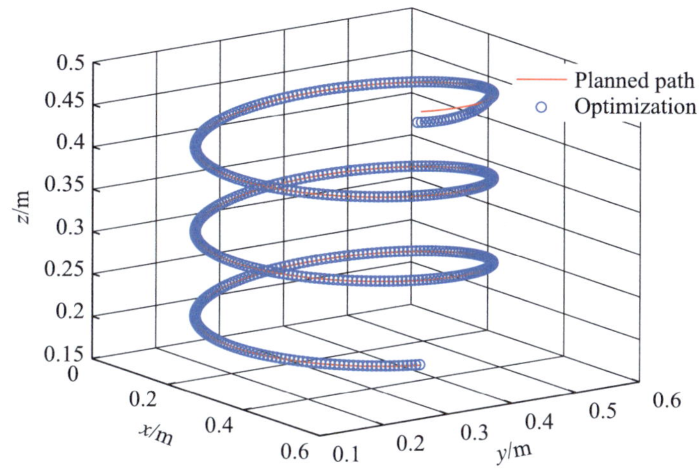

Fig. 5.14 Helical path

the fixed-point arrest experiment of the soft manipulator, the ability of the soft manipulator to achieve the target acquisition according to the task planning path is preliminarily demonstrated. The following will introduce the typical motion of the soft manipulator in turn.

5.4.1 Spiral Winding Movement

When it is not inflated, the position of the soft manipulator is $\boldsymbol{r}_0 = [x_0 \quad y_0 \quad z_0]^{\mathrm{T}} = [0.3070 \quad 0.3120 \quad 0.5418]^{\mathrm{T}}\mathrm{m}$. The helix equation (planned path) in this example is expressed as:

$$\boldsymbol{r} = \begin{bmatrix} x \\ y \\ z \end{bmatrix} = \begin{bmatrix} x_0 + r\cos\theta \\ y_0 + r\sin\theta \\ z_0 - d - \dfrac{\theta}{6\pi}s \end{bmatrix} \tag{5.43}$$

where, $r = 0.2$ m; $d = 0.05$ m; $s = 0.30$ m. The physical meaning of the expression is that the end of the soft manipulator will move from top to bottom along the specified helix, the radius of the helix is 0.20 m, including 3 cycles, and the end of the soft manipulator will drop by 0.10 m in each cycle. The end of the soft manipulator will drop by 0.30m in the whole movement process. The solid red line in Fig. 5.14 shows the planned paths.

Path planning and pressure control strategies of the soft manipulator are adopted to carry out path planning. The problem can be described as: given the motion path, under the control of air pressure, the end of the soft manipulator moves along the given path. The path planning problem can be decomposed into two steps: ①

solving the pressure of the nine air chambers of the soft manipulator according to the specified position of a single point, which is essentially the inverse problem of the soft manipulator kinematics model; ② on the premise of ensuring smooth motion and vibration suppression of the manipulator, the specified motion path is discretized into several position sequences, and a set of pressure sequences of the soft manipulator are solved according to the inverse problem of kinematics. Finally, the nine air chambers of the soft manipulator are inflated according to the time series of the above pressure, so as to realize the purpose that the end of the soft manipulator moves along the specified path.

Firstly, consider the first step, which is to solve the pressure of the nine air chambers of the soft manipulator according to the specified position of a single point. According to the mechanical model of the soft manipulator, the input is the pressure of nine air chambers, and the output is the position vector of the end of the manipulator (there are 3 in total). Given three components of the end position of the manipulator, nine output variables are needed to be solved, which is the multi-solution problem. Here, an optimization algorithm is used to give a feasible solution. The design variable of the optimization problem is the pressure of nine chambers. The optimized objective function is to minimize the distance between the end of the manipulator and the target position. The constraints to be met are nine pressures that cannot exceed a certain threshold to ensure the safety of the structure.

Secondly, considering the second step, according to the specified path, a set of pressure series of the soft manipulator is calculated on the basis of the above single point optimization. As shown in Fig. 5.14, the given motion path is discretized into a sequence of positions. For a given target position vector, a feasible solution can be obtained through the inverse kinematics problem optimization algorithm of the first step, but this solution is not unique. In order to ensure that the pressure changes at the two adjacent discrete positions will not be too large during the continuous motion of the soft manipulator, the initial optimization value of the current step is set as the optimization result of the previous step. After a series of optimization iteration, the final optimization result is shown as the hollow dot in Fig. 5.14.

According to the optimization algorithm, a group of pressure value sequence of nine air chambers is calculated, and the experiment is carried out on the ground system of the soft manipulator. The experimental results are shown in Fig. 5.15. Within each spiral period, the soft manipulator configurations at four different moments were successively captured, and the end of the soft manipulator completed the left-front-right-back circular motion. In addition, according to the spiral equation, the end of the soft manipulator keeps falling in the vertical direction during the whole spiral motion, which is also verified by the ground experiment. For example, the position of the end of the manipulator in the third spiral movement is lower than the position of the end of the manipulator in the first spiral movement.

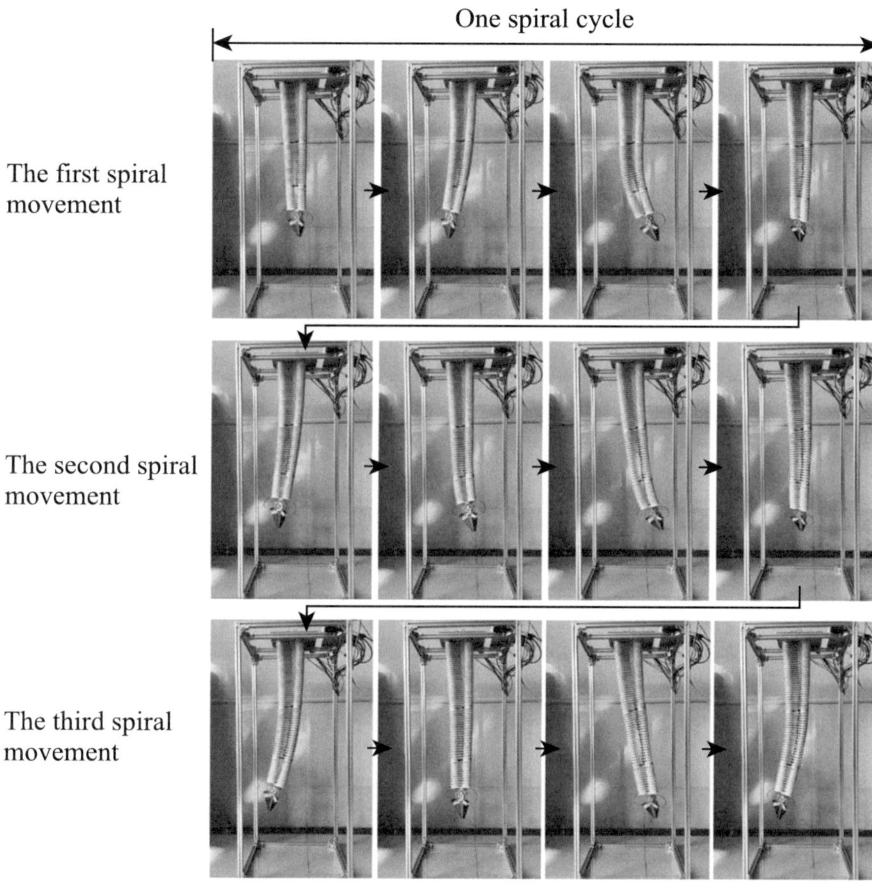

Fig. 5.15 Spiral winding movement

5.4.2 The Four-Leaf Clover Movement

When it is not inflated, the position of the soft manipulator $\boldsymbol{r}_0 = [\,x_0\ y_0\ z_0\,]^{\mathrm{T}} = [\,0.3070\ 0.3120\ 0.5418\,]^{\mathrm{T}}$m, in this example, the four-leaf clover analytic formula can be expressed as:

$$\boldsymbol{r} = \begin{bmatrix} x \\ y \\ z \end{bmatrix} = \begin{bmatrix} \cos(\pi/4) & \sin(\pi/4) & 0 \\ -\sin(\pi/4) & \cos(\pi/4) & 0 \\ 0 & 0 & 1 \end{bmatrix} \begin{bmatrix} r\sin(2\theta)\cos\theta \\ r\sin(2\theta)\sin\theta \\ 0 \end{bmatrix} + \begin{bmatrix} 0 \\ 0 \\ d \end{bmatrix} \tag{5.44}$$

The maximum diameter length of the four-leaf clover $r = 0.50$ m, and $d = -0.20$ m denotes the offset in the vertical direction compared with the initial position where the manipulator is uninflated. The planned path is shown by the red solid line in Fig. 5.16.

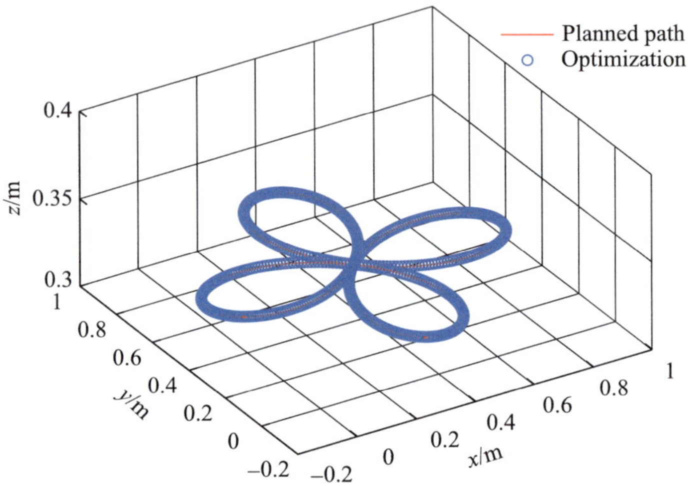

Fig. 5.16 The four-leaf clover path

Path planning and pressure control strategies of the soft manipulator are shown in Fig. 5.17. Under the control of air pressure, the end of the soft manipulator moves along the given path. On the premise of smooth motion and vibration suppression, the specified motion path is discretized into several position sequences, and a set of pressure sequences of the soft manipulator are solved according to the inverse problem of kinematics. The nine chambers of the soft manipulator are inflated according to the time series of the above pressure, so that the end of the manipulator moves along the four-leaf clover path. After a series of optimization iteration, the final optimization result is shown as the hollow blue dot in Fig. 5.16.

According to the optimization algorithm, we calculate the pressure sequence of nine air chambers of the soft manipulator, and the experiment is carried out on the ground system. Experimental results are shown in Fig. 5.17. The manipulator configurations at 4 different moments were captured successively within each lobe period. It can be seen that in each lobe period, the end of the soft manipulator completed four lobe configuration movements, including left-front-right-back. In addition, according to the four-leaf clover equation, the end of the soft manipulator remains unchanged in the vertical direction during the whole motion, which is also verified by ground experiments.

5.4.3 The Pendulum Swings Movement from Side to Side

In this example, the ground experiment of the soft manipulator pendulum swinging motion from side to side is carried out. As shown in Fig. 5.18, according to the layout of the nine air chambers of the soft manipulator, the numbers of the upper, middle

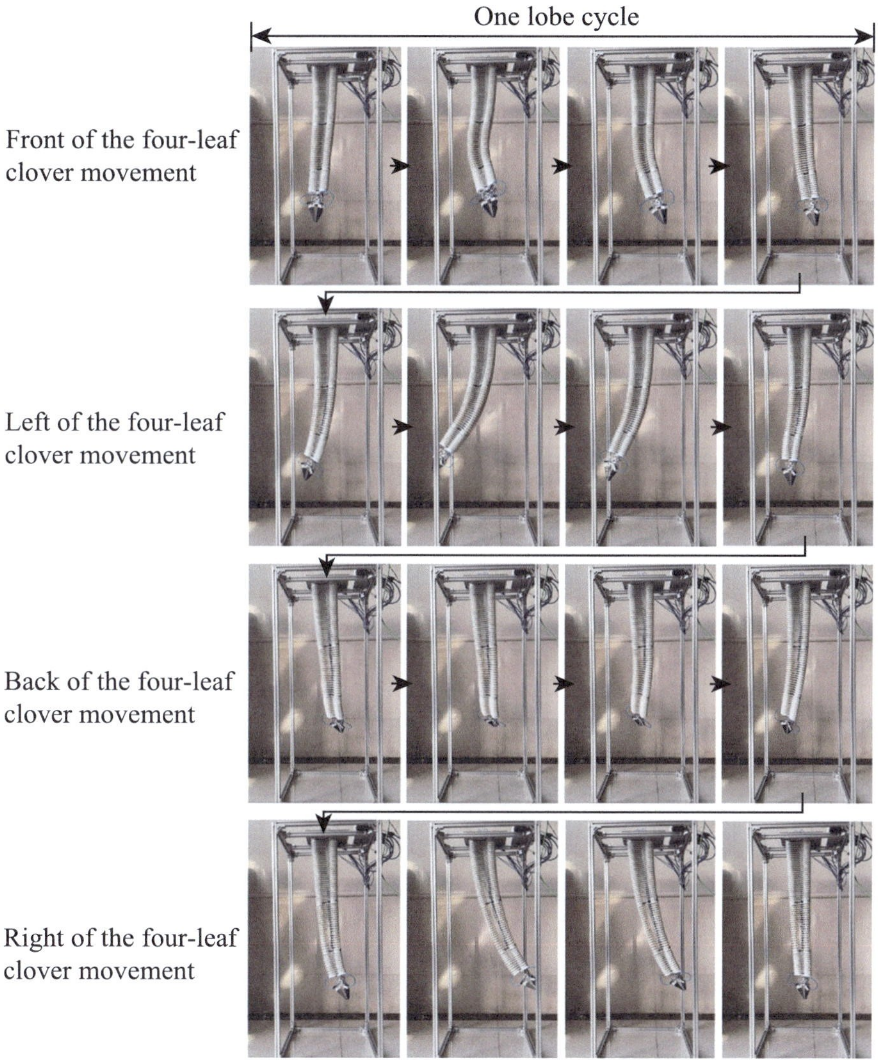

Fig. 5.17 The four-leaf clover movement

and lower three air chambers on the left are successively recorded as 1, 4 and 7, and the numbers of the six air chambers on the right are successively recorded as 2, 3, 5, 6, 8 and 9 from top to bottom. According to the mechanical model of the soft manipulator, the left side 1, 4 and 7 will be inflated at the same time, which will make the soft manipulator bend to the right. Simultaneous inflation of 2, 3, 5, 6, 8, and 9 on the right side will make the manipulator bend to the left.

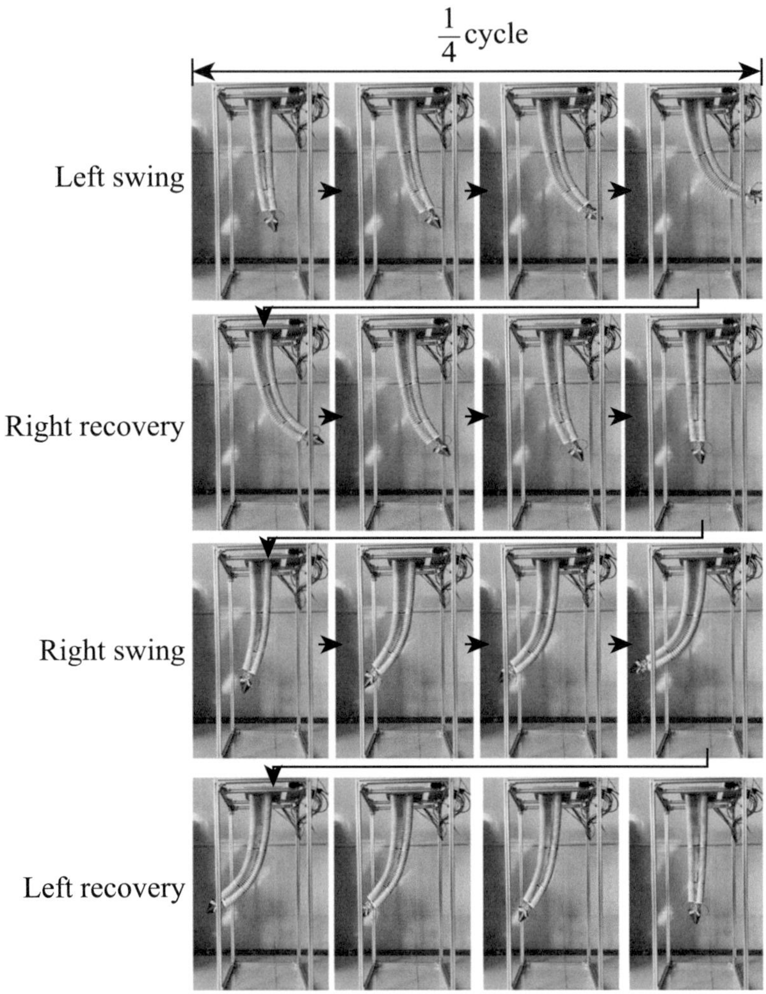

Fig. 5.18 The pendulum swings from side to side

To ensure the safety of the soft arm structure, the upper pressure limit is set to 1200 mbar in this example. The specific filling operation sequence of each air chamber is indicated as follows:

(1) The pressure of left air chambers 1, 4 and 7 are simultaneously linearly increased to 1200 mbar, while the right air chambers 2, 3, 5, 6, 8 and 9 are not operated.
(2) The pressure of left air chambers 1, 4, 7 are simultaneously linearly decreased to 0 mbar, and the right air chambers 2, 3, 5, 6, 8, 9 are not operated.
(3) The pressure of right air chambers 2, 3, 5, 6, 8, 9 are simultaneously linearly increased to 1200 mbar, and the left air chambers 1, 4, 7 are not operated.

(4) The pressure of right air chambers 2, 3, 5, 6, 8, 9 are simultaneously linearly decreased to 0 mbar, and left air chambers 1, 4, 7 are not operated.

During the above inflating and deflating operation, the discrete time interval of each air chamber input is set as $\Delta t = 20\,\text{ms}$. According to the above inflating operation sequence, experiments were carried out on the ground system. The experimental results are shown in Fig. 5.18. It can be seen that under the above inflatable operation sequence, the soft manipulator has successively completed the left swing-right recovery-right swing-left recovery movement.

5.4.4 The Horizontal Plane Swings from Side to Side and Bends Upward Movement

In the several ground experiments above, the soft manipulator was suspended vertically, and ground gravity helped the soft manipulator to return to its vertical configuration, which could not accurately simulate the deformation of the soft manipulator in the absence of gravity in space environment. In order to eliminate the influence of gravity on the inflatable deformation of the soft manipulator, the soft manipulator is placed on a horizontal table in this example. However, the disadvantage of this example is that although the influence of ground gravity on the deformation of the soft manipulator is eliminated, the friction between the manipulator and the table cannot be eliminated during its movement. This problem is also considered as the next research content, that is, to design a reasonable gravity compensation system to ensure that the impact of friction on the soft manipulator is reduced as much as possible while eliminating the impact of ground gravity, for example, the electromagnetic suspension system can be used as the ground gravity compensation.

We fixed the end of the soft manipulator and set the upper pressure to 1200 mbar. The specific inflation operation sequence of each air chamber are as follows:

(1) The pressure of left air chambers 1, 4 and 7 are simultaneously linearly increased to 1200 mbar, while the right air chamber 2, 3, 5, 6, 8 and 9 are not operated.
(2) The pressure of left air chambers 1, 4, 7 are simultaneously linearly decreased to 0 mbar, and the right air chambers 2, 3, 5, 6, 8, 9 are not operated.
(3) The pressure of right air chambers 2, 3, 5, 6, 8, 9 are simultaneously linearly increased to 1200 mbar, and the left air chambers 1, 4, 7 are not operated.
(4) The pressure of right air chambers 2, 3, 5, 6, 8, 9 are simultaneously linearly decreased to 0 mbar, and left air chambers 1, 4, 7 are not operated.

During the above inflating and deflating operation, the discrete time interval of each air chamber input is set as $\Delta t = 20\,\text{ms}$. According to the above inflating operation sequence, experiments were carried out on the ground system of the soft manipulator. The experimental results are shown in Fig. 5.19. It can be seen that under the above inflatable operation sequence, the soft manipulator successively completes

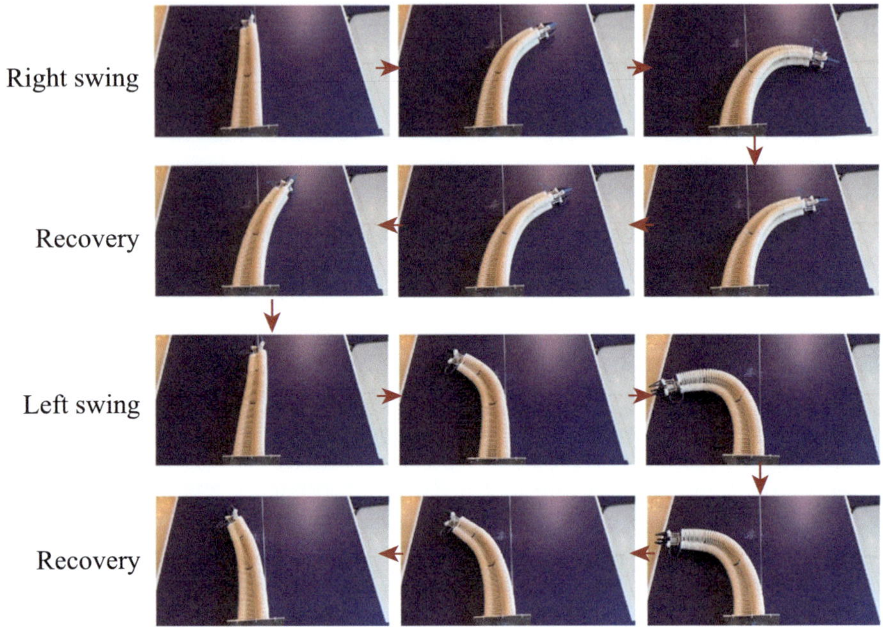

Fig. 5.19 The horizontal plane swings from side to side

the horizontal swing to the right-horizontal swing to the left-horizontal swing to the left-right recovery movement.

In addition, this section also carried out the upward bending experiment of the soft manipulator, in which the manipulator's air chambers 2, 3, 5, 6, 8 and 9 were placed on the table, and the other three air chambers 1, 4 and 7 were placed on the top. We fixed the end of the soft manipulator and set the upper pressure to 1200 mbar. The specific charging operation sequence of each air chamber are as follows: the air pressure of lower air chambers 2, 3, 5, 6, 8 and 9 are simultaneously linearly decreased to 1200 mbar, while the upper air chambers 1, 4 and 7 are not operated. During the above inflating and deflating operation, the discrete time interval of each air chamber input is set as $\Delta t = 20\,\text{ms}$.

According to the above inflating operation sequence, experiments were carried out on the ground system. The experimental results are shown in Fig. 5.20. It can be seen that the soft manipulator completes the upward bending action in turn under the above inflating operation sequence.

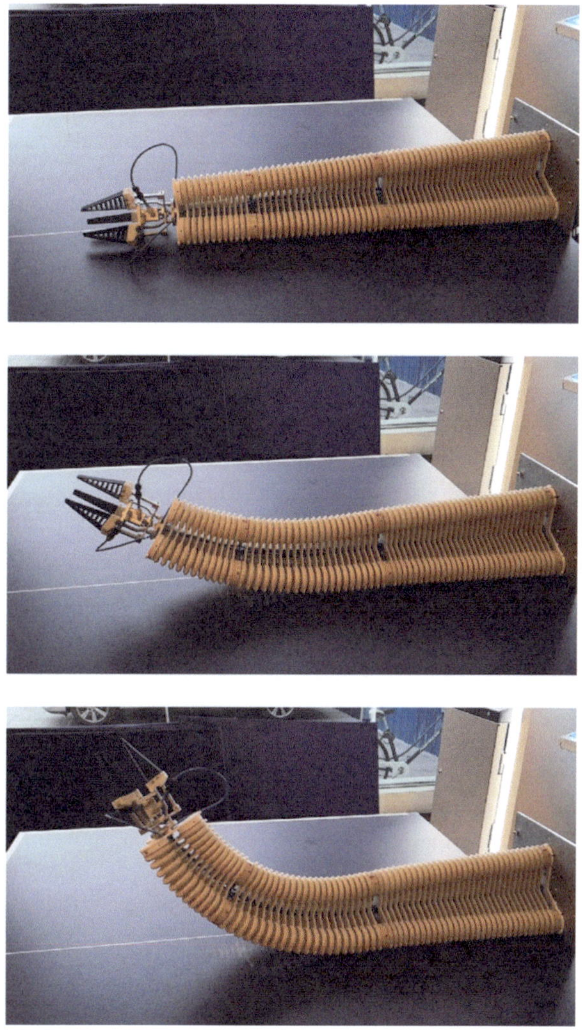

Fig. 5.20 Soft manipulator upward bending experiment

5.4.5 *Fixed Point Movement and Target Capture*

In this section, the fixed-point motion and target capture experiments of the soft manipulator were carried out. The end of the soft manipulator was fixed on the frame, and the upper pressure was set to 1200 mbar. We set the inflating operation sequence of each air chamber and specify the pressure of 1–9 air chambers. The specific gas chamber filling operation sequence is as follows: the pressure of gas chambers 1–9 is linearly increased to the specified pressure at the same time. During the above inflating and deflating operation, the discrete time interval of each air chamber input is set as $\Delta t = 20\,\text{ms}$.

According to the above gas-filling operation sequence, experiments were carried out on the ground system. The experimental results are shown in Fig. 5.21. It can be seen that the soft manipulator completes the fixed-point movement and grasping operation under the above inflatable operation sequence. After that, the end gripper starts, and the pneumatic control system provides two air gripper pressure controls. It controls the internal pressure of air chamber 10 and air chamber 11, so as to realize the positive and negative movement of the cylinder and drive the opening and closing operation of the flexible gripper, as shown in Fig. 5.18. After the soft manipulator reaches the specified target position and the end gripper completes the opening and closing grasp operation, the soft manipulator recovers to its initial position. The specific inflating operation sequence of each air chamber are as follows: the pressure of air chambers 1–9 is linearly decreased to the specified pressure value at the same time. During the above inflating and deflating operation, the discrete time interval of each air chamber input is set as $\Delta t = 20\,\text{ms}$, and the soft manipulator returns to the vertical configuration.

In order to further carry out the target soft capture experiment, the target object (white vase base) is placed in the designated position. According to the optimization algorithm of the soft manipulator, a set of optimal solutions of the pressure of each air chamber are obtained, so that the end of the soft manipulator reaches the specified target position. The specific inflating process of each air chamber is as follows: the gas chamber 1–9 is linearly pressurized to the specified pressure at the same time. During the above inflating and deflating operation, the discrete time interval of each air chamber input is set as $\Delta t = 20\,\text{ms}$.

According to the above inflating operation sequence, experiments were carried out on the ground system. The experimental results are shown in Fig. 5.22. It can be seen that the soft manipulator completes the fixed-point movement and grasping operation under the above inflatable operation sequence. After that, the end gripper starts, and the pneumatic control system provides two air gripper pressure controls to control the internal pressure of air chamber 10 and air chamber 11, so as to realize the positive and negative movement of the cylinder and drive the opening and closing operation of the flexible gripper, as shown in Fig. 5.21. After the soft manipulator reaches the specified target position and the end gripper completes the opening and closing grasp operation, the soft manipulator recovers to its initial position. The specific inflating operation sequence of each air chamber is as follows: the pressure of air chamber 1–9 is linearly decreased to the specified pressure at the same time. During the above inflating and deflating operation, the discrete time interval of each air chamber input is set as $\Delta t = 20\,\text{ms}$, and the soft manipulator finally returns to the vertical configuration.

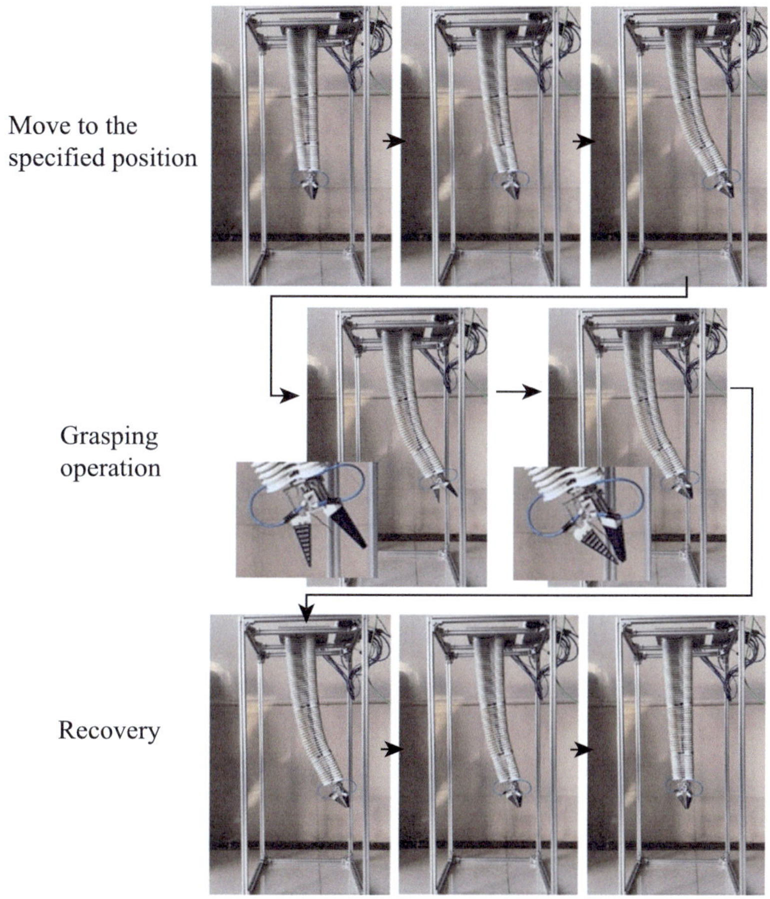

Fig. 5.21 Fixed point motion

5.5 Summary

In this chapter, an improved design is carried out for the scheme of trunk-like soft manipulator, so as to make it conform to the variable cross-section configuration of elephant trunk biological tissue. On this basis, an improved prototype and test system of the trunk-like soft manipulator are developed, and the corresponding path planning and trajectory control are carried out. The main conclusions are as follows:

(1) The variable section configuration design is more in line with the real configuration of the elephant trunk, and the newly developed prototype of the soft manipulator shows more stable and flexible motion characteristics.

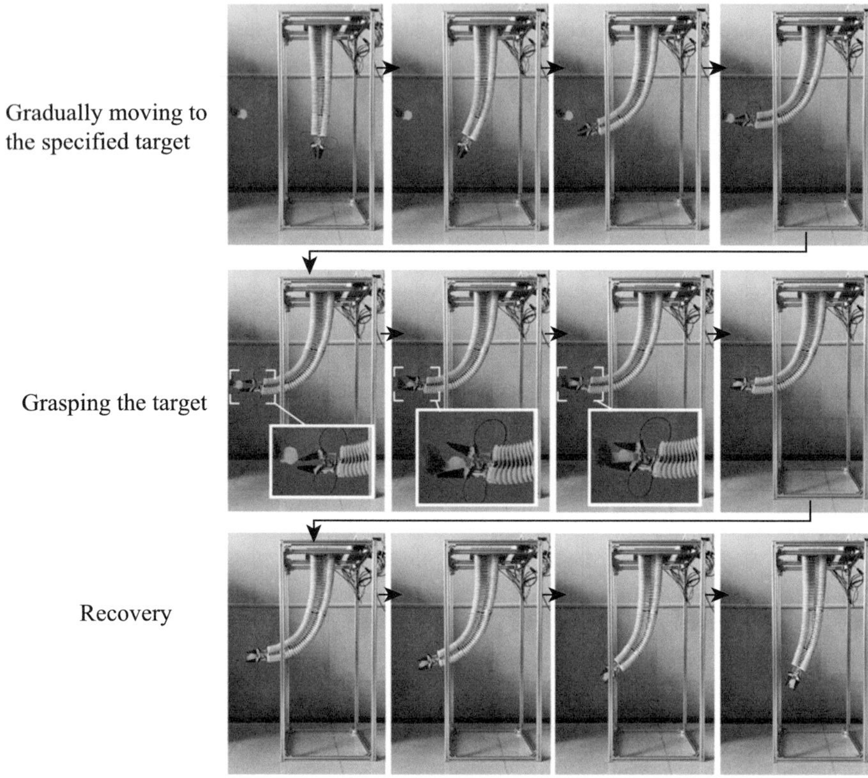

Fig. 5.22 Targeted capture

(2) The simplified model based on the test data can effectively solve the singular problem that may occur in the process of calculation of the inverse kinematics solution, and provide the model basis for path planning and control.
(3) By splitting the specified motion path into several position sequences, and solving a set of pressure sequences of the soft manipulator according to the mechanical model. We then inflate the nine air chambers of the soft manipulator according to the series of time, and achieve more stable motion control of the soft manipulator along the specified path.

References

1. Huang X, Zhu X, Gu G (2022) Kinematic modeling and characterization of soft parallel robots. IEEE Trans Rob 38(6):3792–3806
2. Bamdad M, Bahri MM (2019) Kinematics and manipulability analysis of a highly articulated soft robotic manipulator. Robotica 37(5):868–882

3. Rolf M, Steil JJ (2012) Constant curvature continuum kinematics as fast approximate model for the bionic handling assistant. IEEE/RSJ international conference on intelligent robots and systems. Vilamoura-Algarve, Portugal, pp 3440–3446
4. Fang G, Tian Y, Yang ZX et al (2022) Efficient jacobian-based inverse kinematics with sim-to-real transfer of soft robots by learning. IEEE/ASME Trans Mechatron 27(6):5296–5306
5. Gong Z, Cheng J, Hu K et al (2018) An inverse kinematics method of a soft robotic arm with three-dimensional locomotion for underwater manipulation. IEEE international conference on soft robotics (RoboSoft). Livorno, Italy, pp 516–521
6. Windolf M, Götzen N, Morlock M (2008) Systematic accuracy and precision analysis of video motion capturing systems–exemplified on the vicon-460 system. J Biomech 41(12):2776–2780
7. Mbakop S, Tagne G, Lakhal O, et al (2020) Path planning and control of mobile soft manipulators with obstacle avoidance. In: The 3rd IEEE international conference on soft robotics, pp 64–69
8. Boggs PT, Tolle JW (1995) Sequential quadratic programming. Acta Numer 4:1–51
9. Kuffel P, Kent K, Irwin G (1997) The implementation and effectiveness of linear interpolation within digital simulation. Int J Electr Power Energy Syst 19(4):221–227

Chapter 6
Kinematic Modelling and Experimental Validation of a Foldable Pneumatic Soft Module

If the space manipulator can be folded into a small volume with a light weight before launch, it will definitely have a much greater prospect in space missions. Moreover, this kind of foldable pneumatic manipulator will also have a wide application in future industrial fields, physical human-robot interactions, and many other fields. In confined spaces, where traditional rigid equipment cannot work, the pneumatic soft manipulator shows competent performance. In this chapter, a pneumatic soft manipulator with the foldability to switch between a contraction state and an expanded state is designed and developed. In Sect. 6.1, the design schematic of the soft manipulator is proposed, and the kinematic model of one foldable pneumatic module is developed. In Sect. 6.2, the numerical results, including the shape deformation and the workspace of the soft manipulator, are obtained and analyzed. In Sect. 6.3, the experimental system is designed and developed. The repeatability of the pneumatic module and the prediction accuracy of our proposed model are tested and validated. Finally, the conclusions are given in Sect. 6.4.

6.1 Kinematic Modelling of the Foldable Pneumatic Soft Manipulator

6.1.1 Design Schematic

The foldable pneumatic soft manipulator is designed to have the foldability in its volume [1]. When it is evacuated, the soft manipulator is in a contraction state, which can greatly save the storage space [2]. When it is inflated, the soft manipulator can switch to an expanded state, which can result in a large workspace [3]. This kind of foldability is realized by a structure composed of several foldable pneumatic modules which are all standardized and identical. The design of a foldable pneumatic module is shown in Fig. 6.1.

X. Zhang et al., *Design and Implementation of Soft Robotic Manipulators*,
https://doi.org/10.1007/978-981-95-1657-5_6

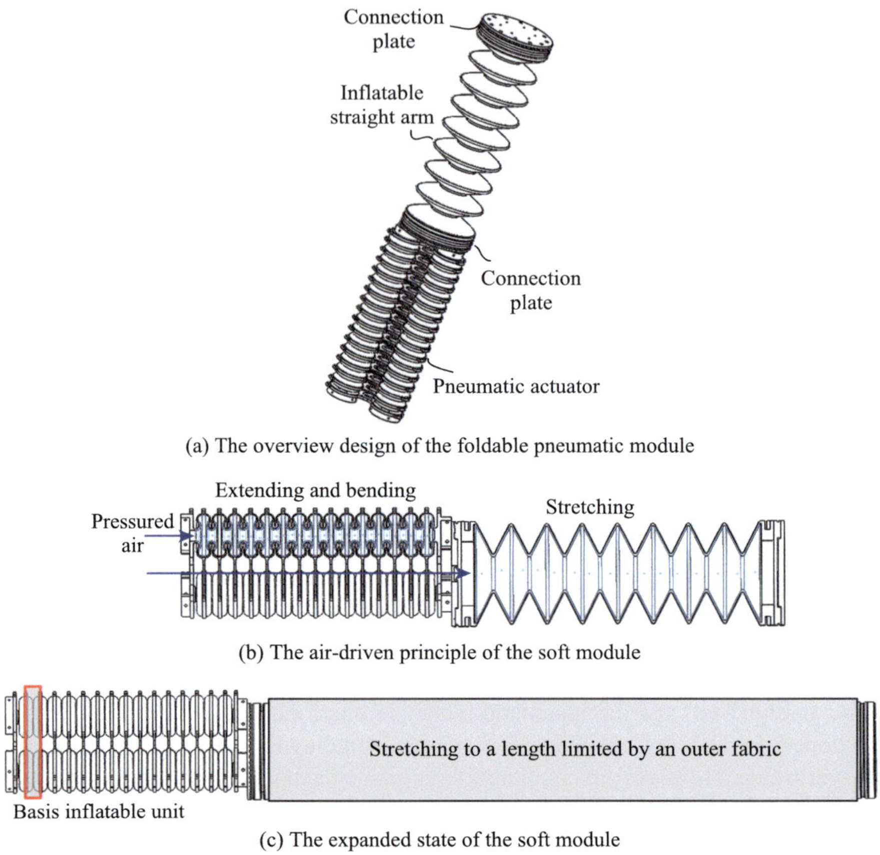

Fig. 6.1 Design of the foldable pneumatic module

In Fig. 6.1a, it is seen that the foldable pneumatic module is composed of a pneumatic actuator, an inflatable straight arm, as well as the connection plate between the two parts. The pneumatic actuator is made up of three parallel and symmetric inflatable bellows, which are connected by a series of rigid constraint plates. The straight arm has a two-level structure: an inner level and an outer level. The inner level is a bellows made of silicone rubber, which is flexible and is responsible for the sealing. The outer level is a cylinder beam made of nylon fabric, which limits the lateral and longitudinal expansion of the inner silicone rubber. In Fig. 6.1b, the air-driven principle of the soft module is shown. The pressured air is inputted into the inflatable bellows and the inflatable straight arm independently. The pneumatic actuator can make the directional motion when the air pressure is different in the three bellows. The inflatable straight arm can switch between a compact state and an expanded state when the pressured air is evacuated or filled. In Fig. 6.1c, the expanded state of the soft module is shown. The total length of the inflatable straight arm is limited by the outer level fabric. The directional movement of the pneumatic actuator is the bending accumulation of a large number of basic inflatable units. By controlling the movement of the pneumatic actuator, the motion control of the soft manipulator can be realized.

6.1.2 Kinematic Model of the Foldable Pneumatic Module

The foldable pneumatic module, which includes a pneumatic actuator and an inflatable straight arm, is the basic unit of the pneumatic soft manipulator. The difficulty in modeling the kinematic performance of the pneumatic module lies in the pneumatic actuator [4]. To develop the kinematic model of the pneumatic actuator, a basic inflatable unit is taken and shown in Fig. 6.2. The inflatable unit is composed of three bellows slices that are arranged in an equilateral triangle configuration, and its upper and lower ends are fixed together by rigid constraint plates. It is assumed that the lower end is fixed and the upper end can move freely.

As is shown in Fig. 6.2, the basic inflatable unit is just a small slice taken from the whole pneumatic actuator. The length of the equilateral triangle configuration is a. When the bellows are not inflated, the bellows slice remains its original length L_0. Once the bellows are inflated, the bellows stretches along the axial direction. The elongation of the manipulator is produced by the extension of the bellow structure. Therefore, the strain of the bellow still agrees with the small deformation assumption.

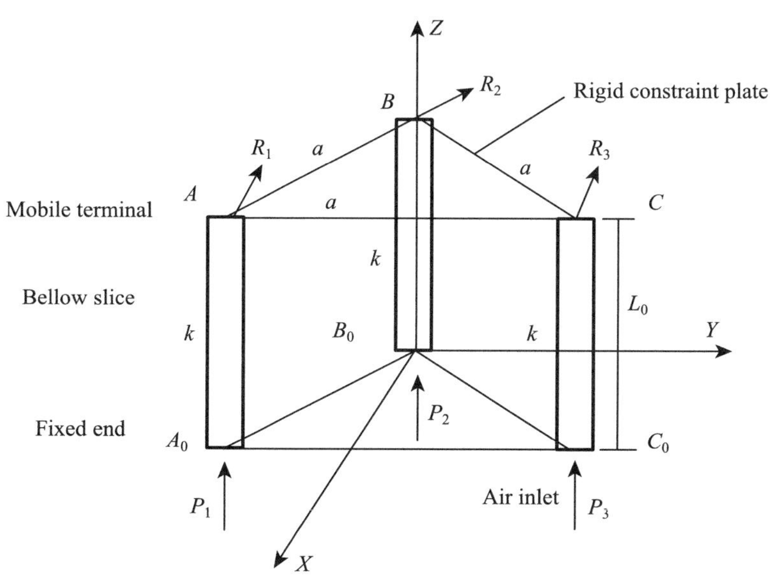

Fig. 6.2 Basic inflatable unit of the pneumatic actuator

Moreover, the axial elongation of the bellows is considered, while the radial inflation is neglected. For the convenience of analysis, the elongated cylindrical bellows is considered as an elastomer. It is assumed that the elastic coefficients of the three bellows are the same, k. The internal pressures of the three bellows are P_1, P_2 and P_3 respectively. The forces produced by the elongation of the three bellows are F_1, F_2 and F_3, and the elongations are ΔL_1, ΔL_2 and ΔL_3 respectively. When the bellows are inflated, the deformations are assumed to be in the elastic range, thus

$$\begin{cases} F_1 = k\Delta L_1 \\ F_2 = k\Delta L_2 \\ F_3 = k\Delta L_3 \end{cases} \tag{6.1}$$

It is assumed that the internal cross-sectional area of the bellows is S. The torque balance of the rigid constraint plate at the upper end gives:

$$\begin{cases} F_1 = k\Delta L_1 = P_1 S \\ F_2 = k\Delta L_2 = P_2 S \\ F_3 = k\Delta L_3 = P_3 S \end{cases} \tag{6.2}$$

After inflation, if there is no upper rigid constraint plate, the lengths of the three bellows are respectively:

$$\begin{cases} L_1 = L_0 + \Delta L_1 = L_0 + \dfrac{P_1 S}{k} \\ L_2 = L_0 + \Delta L_2 = L_0 + \dfrac{P_2 S}{k} \\ L_3 = L_0 + \Delta L_3 = L_0 + \dfrac{P_3 S}{k} \end{cases} \tag{6.3}$$

The mechanical effects of the rigid constraint plates at the upper end are as follows: ① the three points A, B and C at the upper end of the three bellows keep still an equilateral triangle, and the length of the side is a; ② making the normal direction of the upper plates of the three bellows point to the same direction. The forces produced by the upper rigid constraint plates on the three bellows are R_1, R_2 and R_3 respectively, and their directions are located in the plane where the constraint plate is located. Taking the fixed-end plane as the X-Y plane, a cartesian coordinate system is established, as shown in Fig. 6.2. A_0, B_0 and C_0 are the fixed ends of the three bellows respectively. The three points constitute an equilateral triangle, and the length of the side is a. Taking the B_0 point as the origin of the coordinate, the axis x is perpendicular to the A_0C_0 downward direction, the axis y is horizontal to the right, and the axis z is perpendicular to the X-Y plane, forming a right-handed cartesian coordinate system. In this coordinate system, the coordinates of A_0, B_0 and C_0 can be written as:

$$\begin{cases} A_0 : \begin{bmatrix} \frac{\sqrt{3}}{2}a & -\frac{a}{2} & 0 \end{bmatrix} \\ B_0 : \begin{bmatrix} 0 & 0 & 0 \end{bmatrix} \\ C_0 : \begin{bmatrix} \frac{\sqrt{3}}{2}a & \frac{a}{2} & 0 \end{bmatrix} \end{cases} \tag{6.4}$$

In the uninflated state, the coordinates of the A, B and C can be obtained by:

$$\begin{cases} A : \begin{bmatrix} \frac{\sqrt{3}}{2}a & -\frac{a}{2} & L_0 \end{bmatrix} \\ B : \begin{bmatrix} 0 & 0 & L_0 \end{bmatrix} \\ C : \begin{bmatrix} \frac{\sqrt{3}}{2}a & \frac{a}{2} & L_0 \end{bmatrix} \end{cases} \tag{6.5}$$

Considering the effect of R_1, R_2 and R_3, the deflections of the upper end of the three bellows are

$$\begin{cases} v_1 = \frac{R_1 L_1^3}{3E_1 I} = \frac{R_1 L_1^2 S'}{3kI} \\ v_2 = \frac{R_2 L_2^3}{3E_2 I} = \frac{R_2 L_2^2 S'}{3kI} \\ v_3 = \frac{R_3 L_3^3}{3E_3 I} = \frac{R_3 L_3^2 S'}{3kI} \end{cases} \tag{6.6}$$

where, E is the elastic modulus; I is the moment of inertia; EI is the bending stiffness; S' is the cross-sectional area of the wall of the bellows, that is, the effective area under axial tension. For the section of the ring whose outer diameter is R and the inner diameter is r, its moment of inertia is

$$I = \int y^2 \mathrm{d}A = \int_0^{\pi} \left(\frac{R+r}{2}\right)^2 \sin^2\theta \frac{R^2 - r^2}{2} \mathrm{d}\theta = \frac{\pi (R+r)^3 (R-r)}{16} \tag{6.7}$$

Under the action of R_1, R_2 and R_3, the angles generated at the upper end of the three bellows are

$$\begin{cases} \theta_1 = \frac{R_1 L_1^2}{2E_1 I} = \frac{R_1 L_1 S'}{2kI} \\ \theta_2 = \frac{R_2 L_2^2}{2E_2 I} = \frac{R_2 L_2 S'}{2kI} \\ \theta_3 = \frac{R_3 L_3^2}{2E_3 I} = \frac{R_3 L_3 S'}{2kI} \end{cases} \tag{6.8}$$

Considering that the endpoints on the three bellows A, B and C have the same normal direction, it means that:

$$\begin{cases} \theta_1 = \theta_2 \\ \theta_1 = \theta_3 \end{cases} \Rightarrow \begin{cases} R_1 L_1 = R_2 L_2 \\ R_1 L_1 = R_3 L_3 \end{cases} \tag{6.9}$$

The fact that A, B, and C have the same normal orientation means that the projection of the three bellows are parallel in the X-Y plane. Supposed that the angle from the axis x counterclockwise to the projection is α, then the coordinates of the upper endpoints of the three bellows after inflation are respectively as follows:

$$\begin{cases} A' : \left[v_1 \cos\alpha + \frac{\sqrt{3}}{2}a \quad v_1 \sin\alpha - \frac{a}{2} \quad L_1 \right] \\ B' : [v_2 \cos\alpha \quad v_2 \sin\alpha \quad L_2] \\ C' : \left[v_3 \cos\alpha + \frac{\sqrt{3}}{2}a \quad v_3 \sin\alpha + \frac{a}{2} \quad L_3 \right] \end{cases} \tag{6.10}$$

Under the constraint of the upper rigid plates, the distance condition satisfies:

$$L_{A'B'} = a, \quad L_{B'C'} = a, \quad L_{C'A'} = a \tag{6.11}$$

To simplify the equation, it is supposed that:

$$R_1 L_1 = R_2 L_2 = R_3 L_3 = W \geqslant 0, \quad \frac{S'}{3kI} = M > 0 \tag{6.12}$$

It is assumed that the elongations of the three bellows are all greater than zero and less than ΔL, then there is $0 \leqslant \Delta L_1,\ \Delta L_2,\ \Delta L_3 \leqslant \Delta L$. It can be proved that the established condition of the expression $3a^2 - 3(\Delta L_1 - \Delta L_3)^2 - (2\Delta L_2 - \Delta L_3 - \Delta L_1)^2 \geqslant 0$ is $\Delta L/a \leqslant \sqrt{3}/2$, which means that the ratio of the bellow elongation to the bellow diameter is less than $\sqrt{3}/2$. Considering that the elongation of each section of the pneumatic actuator is small, the condition is satisfied in actual motion.

When $\Delta L_1 \neq \Delta L_3$, it can be gotten:

$$W = \frac{a \sin\alpha}{M(\Delta L_1 - \Delta L_3)} \pm \frac{1}{M} \operatorname{sign}(\Delta L_1 - \Delta L_3) \cdot \sqrt{\frac{3a^2 - 3(\Delta L_1 - \Delta L_3)^2 - (2\Delta L_2 - \Delta L_3 - \Delta L_1)^2}{3(\Delta L_1 - \Delta L_3)^2 + (2\Delta L_2 - \Delta L_3 - \Delta L_1)^2}} \tag{6.13}$$

When $\Delta L_1 = \Delta L_3 \neq \Delta L_2$, the W can be expressed as:

$$W = \frac{-\sqrt{3}a \cos\alpha}{M(\Delta L_1 - 2\Delta L_2 + \Delta L_3)} \pm \frac{1}{M} \operatorname{sign}(\Delta L_1 - 2\Delta L_2 + \Delta L_3) \cdot \sqrt{\frac{3a^2 - 3(\Delta L_1 - \Delta L_3)^2 - (2\Delta L_2 - \Delta L_3 - \Delta L_1)^2}{3(\Delta L_1 - \Delta L_3)^2 + (2\Delta L_2 - \Delta L_3 - \Delta L_1)^2}} \tag{6.14}$$

The root W of the equation contains two positive solutions, the smaller one of which is the actual solution to the equation. This is because the smaller one corresponds to a small deflection which agrees with the physical truth. The function of the rigid constraint plates is to ensure that the three central points of the bellows keep an equilateral triangle with a side length of a. All the flexible bellows are identical and have the same inside diameter, outside diameter, length and elasticity coefficient.

According to the symmetry, the motion of each bellow unit is the same with respect to its own coordinate system, as shown in Fig. 6.2. Transfer the coordinate B_0-xyz to a new origin point B, and the definition of the new coordinate $x_1y_1z_1$ in the plane ABC is the same with the coordinate xyz. Then the transferring method from the coordinate $x_0y_0z_0$ to the coordinate $x_1y_1z_1$ without considering the origin shift can be written as:

$$\begin{bmatrix} x_1 \\ y_1 \\ z_1 \end{bmatrix} = \boldsymbol{H} \begin{bmatrix} x_0 \\ y_0 \\ z_0 \end{bmatrix} = \boldsymbol{M}_3(-\alpha)\,\boldsymbol{M}_2(\theta)\,\boldsymbol{M}_3(\alpha) \begin{bmatrix} x_0 \\ y_0 \\ z_0 \end{bmatrix} \tag{6.15}$$

where $\boldsymbol{H}$ is the transferring matrix, thus

$$\begin{bmatrix} x_0 \\ y_0 \\ z_0 \end{bmatrix} = \boldsymbol{H}^{\mathrm{T}} \begin{bmatrix} x_1 \\ y_1 \\ z_1 \end{bmatrix} \tag{6.16}$$

When the origin shift $\overrightarrow{B_0B_1}$ is considered, the transferring method from the coordinate $x_0y_0z_0$ to the coordinate $x_1y_1z_1$ can be written as:

$$\begin{bmatrix} x_0 \\ y_0 \\ z_0 \end{bmatrix} = \boldsymbol{H}^{\mathrm{T}} \begin{bmatrix} x_1 \\ y_1 \\ z_1 \end{bmatrix} + \overrightarrow{B_0B_1} \tag{6.17}$$

where $\overrightarrow{B_0B_1}$ can be obtained by combining the Eqs. (6.4) and (6.10).

$$\overrightarrow{B_0B_1} = [v_2\cos\alpha \quad v_2\sin\alpha \quad L_2]^{\mathrm{T}} \tag{6.18}$$

By using this transferring method, the motion of the whole pneumatic actuator can be described. Assumed that the total number of the bellow slices is N in the pneumatic actuator, the coordinate of the top plane center of the pneumatic actuator is $\boldsymbol{r}^*$, then the coordinate of $\boldsymbol{r}^*$ relative to the fixed coordinate system $x_0y_0z_0$ can be expressed as:

$$\boldsymbol{r}_N = \left(\boldsymbol{H}^{\mathrm{T}}\right)^{N-1}\boldsymbol{r}^* + \left[\left(\boldsymbol{H}^{\mathrm{T}}\right)^{N-2} + \left(\boldsymbol{H}^{\mathrm{T}}\right)^{N-3} + \cdots + \left(\boldsymbol{H}^{\mathrm{T}}\right)^{1} + \left(\boldsymbol{H}^{\mathrm{T}}\right)^{0}\right]\overrightarrow{B_0B_1} \tag{6.19}$$

Using the above equation, the directional motion of the pneumatic actuator can be depicted. The inflatable straight arm is fixed at the end of the pneumatic actuator,

Table 6.1 Model parameters of pneumatic actuator and straight arm

Parameter	L_0	a	N	L_{empty}
Value	0.02 m	0.1 m	16	0.2 m
Parameter	d_{out}	d_{in}	T	L_{full}
Value	0.05 m	0.0488 m	0.02 m	0.442 m

and the elongation of the straight arm is in the tangent direction of the end of the pneumatic actuator. The stiffness of the straight arm with a certain air pressure can reach to a much higher level than the pneumatic actuator, thus the straight arm can be simplified as a straight line in the kinematic modelling process. Then the coordinate of the free top of the straight arm $\boldsymbol{r}_{TS}$ can be obtained by:

$$\boldsymbol{r}_{TS} = \left(\boldsymbol{H}^{T}\right)^{N}\begin{bmatrix} 0 \\ 0 \\ L_S \end{bmatrix} + \left[\left(\boldsymbol{H}^{T}\right)^{N-1} + \left(\boldsymbol{H}^{T}\right)^{N-2} + \cdots + \left(\boldsymbol{H}^{T}\right)^{1} + \left(\boldsymbol{H}^{T}\right)^{0}\right]\overrightarrow{B_0B_1} \tag{6.20}$$

where, L_S is the length of the straight arm. Until now, the kinematic model of the pneumatic soft module has been developed. The shape deformation and the workspace of the pneumatic soft module can be described and analyzed based on this kinematic model.

6.2 Numerical Calculation and Analysis

6.2.1 Model Parameters

In order to evaluate the prediction results of the kinematic model of the pneumatic module, the programming is coded in MATLAB [5], which is a multi-paradigm numerical computing environment developed by MathWorks, an American privately held software corporation. The model parameters, as well as their values used in the numerical calculation, are listed in Table 6.1. L_0 is the original length of each bending unit, a is the distance between each bending unit, N is the number of bending units, d_{out} is the outer diameter of the bending unit, d_{in} is the inner diameter of the bending unit, T is the height of the bending unit, L_{empty} is the length of the straight arm in the empty state, and L_{full} is the length of the inflated state. The specific parameters are adjusted according to each calculation example.

In the calculation process, the elastic coefficient K of the pneumatic actuator has a significant effect on the calculation results. Due to the complex structure of the pneumatic actuator, there is no theoretical formula to directly give the K value, so we need to measure it through experiment. The experimental principle is similar to the spring elastic coefficient measurement method. By measuring the relationship between the elongation and the pressure, the elastic coefficient K value of the pneumatic actuator is obtained by fitting with considering the influence of gravity. Finally, in the calculation process, the elastic coefficient K value of the pneumatic actuator is given as 23,000 N/m.

6.2.2 Shape Deformation of the Pneumatic Module

In this calculation, the internal pressure of the straight arm is fixed at 0.025 MPa. The loading air pressure of the pneumatic actuator is taken from 0.02 to 0.1 MPa, with an interval of 0.02 MPa. Firstly, we inflate one bellows; The other two bellows are not inflated. Then we inflate two bellows with the same pressure, while the other one is not inflated. The numerical results are shown in Figs. 6.3 and 6.4.

It can be seen in Figs. 6.3a and 6.4a that the shape deformation increases with the increase of the internal pressure with both one bellows and two bellows inflated. The line of the straight arm is just along the tangent direction of the end point of the pneumatic actuator. It can be seen in Figs. 6.3b and 6.4b that the projection of the numerical results in the X-Y plane is a straight line, and the projections in these two situations are almost symmetric about the Y-axis. This phenomenon is consistent with the physical law, indicating that there is no singularity and disturbance occurring during the deformation process predicted by the theoretical model. From the analysis above, it is verified that the numerical results of the theoretical model are reasonable. However, the accuracy of the shape deformation predicted by the proposed model still needs to be validated through experiments.

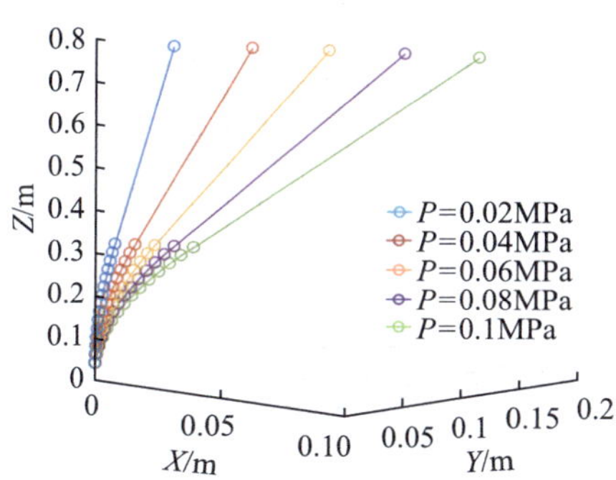

(a) Shape deformation

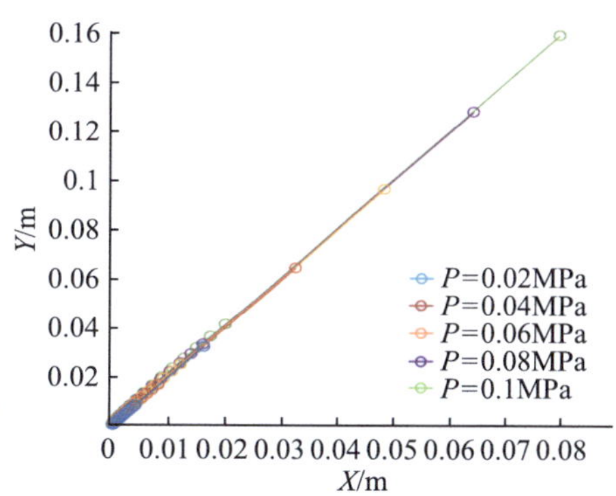

(b) Projection in the X-Y plane

Fig. 6.3 Space distribution of the soft manipulator with one bellows inflated

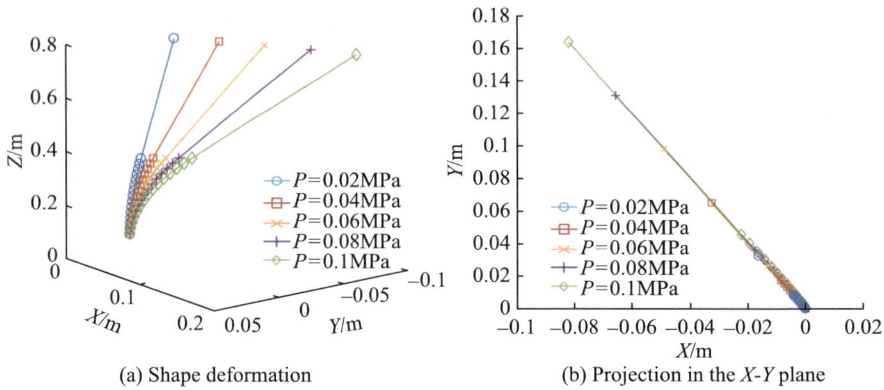

Fig. 6.4 Space distribution of the soft manipulator with two bellows inflated

6.2.3 *Workspace of the Pneumatic Module*

The workspace is an important index to measure the performance of the manipulator [6]. According to the requirements for different mission scenarios, a design scheme of combination structure is proposed. With the combination of the pneumatic actuator and the straight arm, the workspace of the manipulator can be increased efficiently by changing the length of the straight arm. If the manipulator only consists of a pneumatic actuator, the overall weight of the manipulator will be greatly increased. However, the combined structure of the flexible pneumatic actuator and lightweight straight arm can not only reduce the overall weight of the manipulator, but also greatly expands the workspace of the manipulator. In order to compare the workspace with or without considering the straight arm, two examples are numerically calculated based on the proposed model of the soft manipulator. In these two examples, the three bellows are inflated from 0.01 to 0.1 MPa randomly, and the maximum workspaces with or without considering the straight arm are obtained and shown in Fig. 6.5.

In Fig. 6.5, the arc surface marked with blue stars represents the workspace of the combination structure of the straight arm and the pneumatic actuator, while the arc surface marked with red circles represents the workspace of the pneumatic actuator alone. Figure 6.5a shows the comparison of the two kinds of workspace in 3D view; Fig. 6.5b shows the projection of the two workspaces in X-Y plane. From these comparisons in different views, it can be clearly known that the straight arm extends the workspace of the soft manipulator largely. It can be seen in Fig. 6.5a that the projection of the workspace in the X-Y plane is almost hexagonal. The boundary of the hexagon is limited by the maximum internal pressures of the three bellows, which are limited in 0.1 MPa in these two numerical examples. The air pressures of the six vertices of the hexagon are shown in Table 6.2.

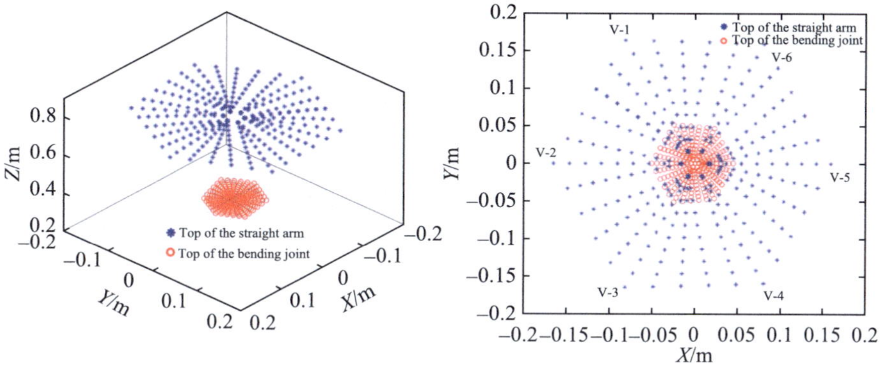

(a) Workspace distribution

(b) Projection of the workspace in the X-Y plane

Fig. 6.5 Workspace of the end point of the soft manipulator

Table 6.2 The air pressures of the six vertices (unit: MPa)

Air pressure	Vertices					
	V-1	V-2	V-3	V-4	V-5	V-6
Pressure in bellow-1	0.1	0.1	0.1	0	0	0
Pressure in bellow-2	0.1	0	0	0	0.1	0.1
Pressure in bellow-3	0	0	0.1	0.1	0.1	0

6.3 Experiment Design and Validation

6.3.1 Experiment Design

The prototype of the foldable pneumatic module is made up of a pneumatic actuator and a straight arm. The pneumatic actuator is fabricated by selective laser sintering (SLS) 3D printing technology [7] with nylon composite powers. It consists of three independent inflatable bellows. The two sides of the pneumatic actuator are fixed by two rigid constraint planes to ensure the three bellows remain in the same plane. The straight arm consists of an inner soft retractable bellows and an outer skin. The outer skin is made of soft fiber fabric which can protect the inner soft bellows and limit the elongation of the inner bellows. The straight arm is in a contracted state when it is not inflated; after inflation, the inner bellows stretches axially to a fixed length limited by the outer skin, as shown in Fig. 6.6.

In Fig. 6.6a and b, the contraction state and the expanded state of the pneumatic soft module are shown respectively. In Fig. 6.6c, the diagram of the pneumatic control system is shown. The pneumatic control system includes a FESTO motion terminal

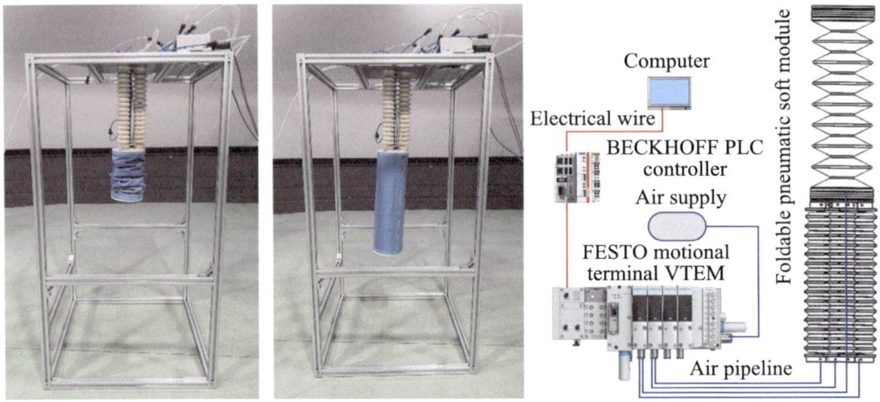

(a) Contraction state (b) Expanded state (c) Pneumatic control diagram

Fig. 6.6 Prototype of the foldable pneumatic soft module

VETM, a BECKHOFF PLC controller, a computer, an air supply, electrical wires, and air pipelines. VTEM is able to control eight pipelines at a proportional pressure with four piezo pilot valves integrated in a valve terminal. The CX9020 BECKHOFF PLC controller is used to offer multi-pipelines air pressure control for the VTEM. The experimental system consists of a pneumatic control system, a measurement system, and a pneumatic soft module, as shown in Fig. 6.7a. The measurement system includes an electronic total station and a force gauge system. On the soft manipulator, nine observation points are set and recorded in order to obtain the shape deformation of the total soft module. Five observation points are on the pneumatic actuator with equal distance, while four observation points are on the straight arm, as shown in Fig. 6.7b. The positions of the nine observation points are recorded by using the electronic total station. The position and status of the pneumatic actuator can be obtained by measuring the position of the five observation points. The number 6 observation point is set at the junction of the pneumatic actuator and the straight arm, which is the starting point of the straight arm. Two observation points are set in the middle section of the straight arm, and one observation point is arranged at the end of the straight arm. The position and status of the straight arm can be obtained by measuring the position of the four observation points.

6.3.2 *Repeatability Test of the Pneumatic Module*

In order to verify the repeated positioning accuracy of the soft manipulator, repeated experiments are taken in different working conditions. Experiments under the same working condition are taken three times, and the coordinates of the observation points are recorded. By comparing the difference of the coordinates of the same observation

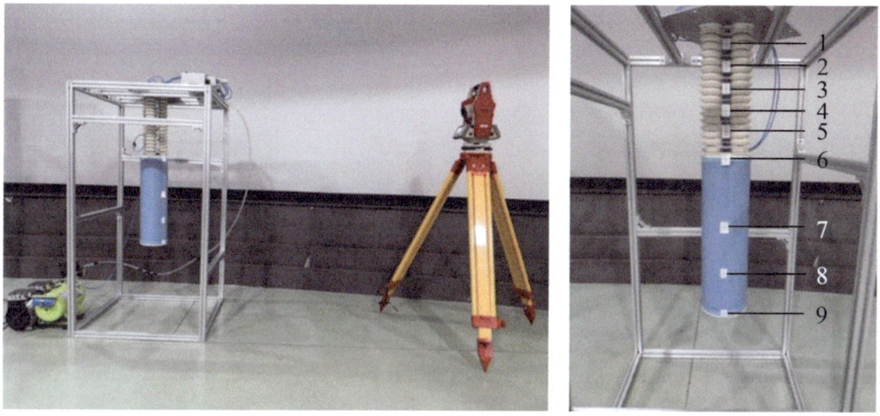

(a) Measurement system (b) Observation points

Fig. 6.7 Experiment system of one pneumatic module

point in three experiments, the repeated positioning accuracy of the pneumatic soft module is tested. Experiments are taken in two conditions: one bellows inflated or two bellows inflated simultaneously.

6.3.2.1 Repetition Experiment with One Bellows Inflated

Firstly, one bellows is inflated, and the internal pressure is set from 0.02 to 0.1 MPa, with 0.02 MPa as the interval. The coordinates of the observation points are recorded, respectively. In each test, the air in the bellows is vented and then inflated. The experimental results are shown in Fig. 6.8.

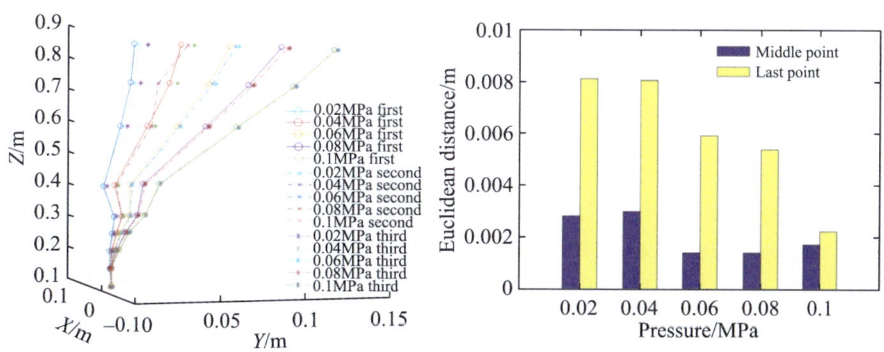

(a) Shape deformation (b) Maximum position error

Fig. 6.8 Repetition experiment with one bellows inflated

In Fig. 6.8a, the curves agree well with each other at the same internal pressure in the three experiments. As the internal pressure increases, the curves are almost completely coincided. It can be seen in Fig. 6.8a that there is a slight difference between the three curves at a pressure of 0.02 MPa, as the internal pressure increases and the difference becomes smaller. The curves are almost completely coincident at the internal pressure of 0.1 MPa. This indicates that the stability of the soft module improves when the pressure increases. As shown in Fig. 6.7b, the observation point number 6 is the end point of the pneumatic actuator, and the observation point number 9 is the end point of the straight arm. These two points are taken as calibration points to obtain the maximum position errors in the three measurements. As can be seen in Fig. 6.8b, the maximum error of the repeated positioning accuracy of the two calibration points is no more than 8 mm, and the repeated positioning accuracy improves with the increase of the internal pressure. When the internal pressure increases to 0.1 MPa, the position error can be limited within 2 mm.

6.3.2.2 Repeatability Experiment When Two Bellows Are Inflated

We then inflate two bellows with the same internal pressure after the pneumatic soft module restores to its original state, and the internal pressure is set from 0.02 to 0.1 MPa, with 0.02 MPa as the interval. The observation process is the same as the experiment with one bellows inflated. The experiment results are shown in Fig. 6.9.

In Fig. 6.9a, the curves almost totally agree with each other at the same internal pressure in the three experiments. The difference between the curves in the same internal pressure condition is very small, no matter the internal pressure is low or high. Similarly, observation points 6 and 9 are taken as calibration points. It can be seen in Fig. 6.9b, the maximum position error of the number 9 observation point is about 6 mm; The maximum position error of the number 6 observation point is about 2 mm. It can be seen from the comparison between Figs. 6.8b and 6.9b that the repeated positioning accuracy improves with the increasing of the internal pressure in the scenario where one bellows is inflated, while the repeated positioning accuracy is stable in the scenario where two bellows are inflated. This is because, when two bellows are inflated at the same time, the bending force generated by the pneumatic actuator is greater than that of one bellows and the stability of the pneumatic soft module is also better. This can be seen in Figs. 6.8 and 6.9, where the maximum error of the repeatability experiment is 8 mm, which is less than 1.5% relative to the total length of the pneumatic soft module (762 mm).

6.3.3 *Validation of the Model Prediction Accuracy*

In order to verify the accuracy of the shape deformation predicted by the proposed model, experiments are taken in different working conditions and the numerical results are compared with the experimental results. The coordinates of the observation

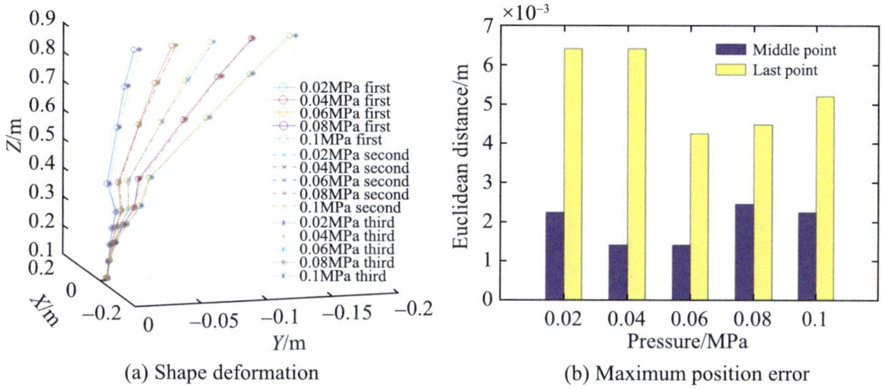

Fig. 6.9 Repeatability experiment when two bellows are inflated

points are recorded in the experiments in two conditions: one bellows inflated or two bellows inflated simultaneously. The coordinates of the observation points are projected into the Y-Z plane and the X-Y plane to see the difference between the modelling prediction and the experiment measurement clearly.

6.3.3.1 Experiment with One Bellows Inflated

First, one bellows is inflated, and the internal pressure is set from 0.02 to 0.1 MPa, with 0.02 MPa as the interval. The coordinates of the observation points are recorded respectively. In this theoretical prediction, the K value is taken as 23,000 N/m; The length of the straight arm is 0.442 m. Observation points 6 and 9 in Fig. 6.7b are taken as calibration points to obtain the errors between the theoretical prediction results and experiment results. The coordinates of these two points are extracted at different internal pressure and compared with the prediction results. The theoretical prediction results and experiment results are compared in Fig. 6.10.

In Fig. 6.10a, the shape deformations of the soft module obtained by both experiment and numerical prediction are shown. It can be seen that the distribution of experiment points is almost around the theoretical prediction. When the curves and points are projected into the X-Y plane, the theoretical results merge into a straight line and the experiment points stay around the straight line, as shown in Fig. 6.10b. This indicates that the pneumatic soft module deforms stably in a single direction under the condition of one bellows inflated, and there is no obvious deviation occurring in the deforming process, which reflects the well stability of the pneumatic soft module. In Fig. 6.10c, the prediction error of the theoretical model relative to the experiment is shown. It can be seen that the prediction error at both the middle point and the last point reduces with the increasing of the air pressure in the bellow. The maximum at the last point is 33 mm when the internal pressure is 0.02 MPa, and the minimum is 23 mm at the internal pressure of 0.1 MPa. The maximum and minimum

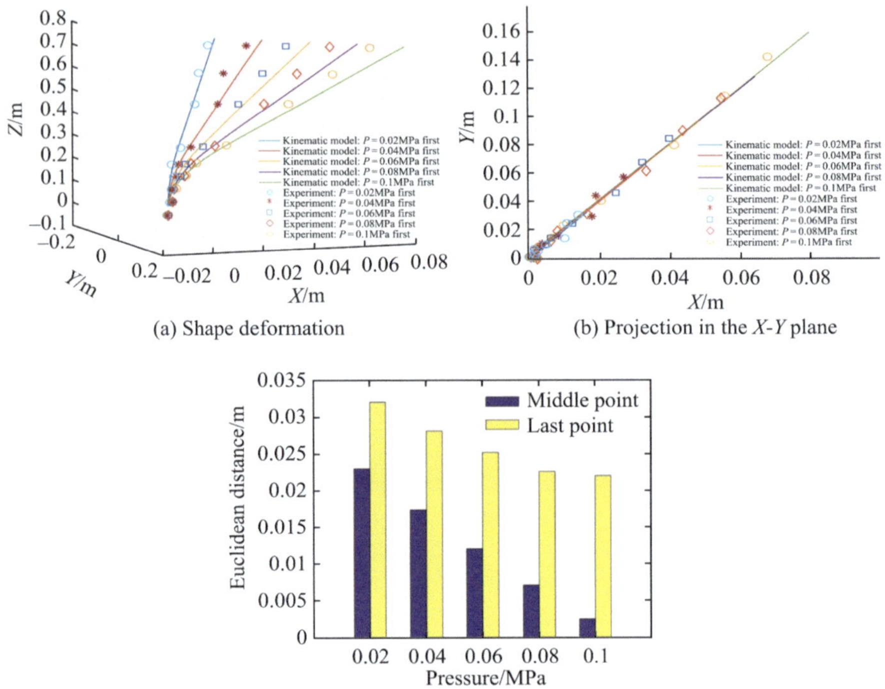

(a) Shape deformation (b) Projection in the *X*-*Y* plane

(c) Error between prediction and experiment results of calibration points

Fig. 6.10 Comparison between prediction and experiment results of one bellows

at the middle point is respectively 24 mm and 2.5 mm. This is because the bending stiffness of the pneumatic joint is relatively small when the internal pressure is at a low level; Thus, the bending amplitude of the pneumatic soft module is greatly influenced by gravity. With the increasing of the internal pressure, the bending stiffness increases; Thus, the influence of gravity on the bending amplitude of the pneumatic soft module reduces relatively [8].

6.3.3.2 Experiment with Two Bellows Inflated

Inflating two bellows with the same internal pressure after the pneumatic soft module restores to its original state, and the internal pressure is set from 0.02 to 0.1 MPa, with 0.02 MPa as the interval. The process is the same as the experiment with one bellows inflated. The experiment results and prediction results are shown in Fig. 6.11.

In Fig. 6.11a, the shape deformations of the soft module obtained by both experiment and numerical prediction are shown. It can be seen that the distribution of experiment points is almost around the theoretical prediction. When the curves and points are projected into the *X*-*Y* plane, the theoretical results merge into a straight line and the experiment points stay around the straight line, as shown in Fig. 6.11b.

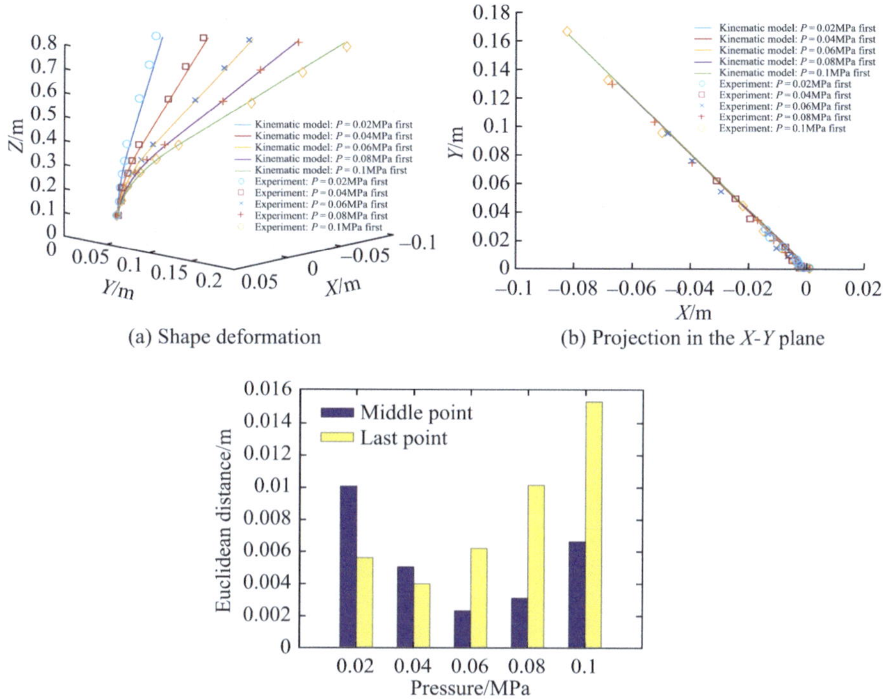

(a) Shape deformation

(b) Projection in the X-Y plane

(c) Error between prediction and experiment results of calibration points

Fig. 6.11 Comparison between prediction and experiment results of two bellows

Comparing Fig. 6.11a with Fig. 6.10a, it can be seen that the agreement between the theoretical prediction and the experiment improves a great deal. This agreement can also be found in their projection to the X-Y plane. It indicates that the stability of the pneumatic soft module and the numerical prediction accuracy are far better with two bellows inflated than with only one bellow inflated. In Fig. 6.11c, the prediction error of the theoretical model relative to the experiment is shown. It can be seen that the minimum prediction error at the middle point is 2 mm, appearing at the internal pressure of 0.06 MPa. When the pressure is lower than 0.06 MPa, the prediction error reduces with the increase of the pressure; When the pressure is higher than 0.06 MPa, the prediction error increases with the increase of the pressure, and the maximum of the error at the last point is less than 16 mm, which is just half of that in the one-bellows condition. The reason lies in that the bending stiffness of the pneumatic joint is strengthened when two bellows are inflated. As a consequence, the stability of the pneumatic soft module improves a great deal and the influence of gravity is relatively small. However, due to the limitations of 3D printing technology, the internal structure of each bellows may not be totally identical. When the internal pressure is lower than 0.06 MPa, the elongation and deformation of the two inflatable bellows are almost identical. When the pressure is higher than 0.06 MPa, the difference in the internal structure of the two bellows is enlarged, leading to an increase of the prediction error.

6.4 Summary

In this chapter, a design of a foldable pneumatic soft manipulator composed of pneumatic actuators and inflatable straight arms was presented. The pneumatic actuators were driven by pressured air and the inflatable straight arms could switch between the contraction state and the extended state. The foldability of the soft manipulator was realized by the inflatable straight arms. Based on this design, the kinematic model of one foldable pneumatic module was proposed. From the numerical results of the shape deformation and the workspace, it can be determined that the prediction of the proposed model was reasonable. In order to validate the design and the theoretical model, an experiment system including a prototype of the pneumatic soft module, a pneumatic control system, and an electronic total station was developed. Through the experiment, the repeatability of the pneumatic soft module was tested and the model prediction accuracy was validated. It can be determined that the performance of the pneumatic soft module was stable, especially in the condition of two bellows inflated. The prediction of the shape deformation by the proposed model was validated by the experimental results, although the error still existed obviously under some conditions. Based on the modelling method presented in this chapter, the kinematic model of the soft manipulator composed of several pneumatic modules could be developed. Due to the advantage of the foldability, the foldable pneumatic soft manipulator will show a wide application prospect in the aerospace field in the future. In our future work, a soft manipulator composed of three pneumatic modules will be designed and developed. The experiment will be conducted based on the air floating platform, which can provide a free moving environment for the planar motion of the soft manipulator.

References

1. Liu Z, Zhang X, Wang J et al (2022) Design, modelling and implementation of a foldable pneumatic soft manipulator. Micrograv Sci Technol 34(3):44
2. Jiang Q, Xu F (2020) Design and motion analysis of adjustable pneumatic soft manipulator for grasping objects. IEEE Access 8:191920–191929
3. Shiva A, Stilli A, Noh Y et al (2016) Tendon-based stiffening for a pneumatically sctuated soft manipulator. IEEE Robot Autom Lett 1(2):632–637
4. Bishop-Moser J, Kota S (2015) Design and modeling of generalized fiber-reinforced pneumatic soft actuators. IEEE Trans Robot 31(3):536–545
5. Gilat A (2017) MATLAB: an introduction with applications. Wiley
6. Amehri W, Zheng G, Kruszewski A (2021) Workspace boundary estimation for soft manipulators using a continuation approach. IEEE Robot Autom Lett 6(4):7169–7176

7. Wallin TJ, Pikul J, Shepherd RF (2018) 3D printing of soft robotic systems. Nat Rev Mater 3(6):84–100
8. Peng X, Zhang N, Ge L et al (2019) Dimension optimization of pneumatically actuated soft continuum manipulators. In: The 2nd IEEE international conference on soft robotics (RoboSoft). Seoul, Korea (South), pp 13–18

Chapter 7
Design, Modelling and Implementation of a Foldable Pneumatic Soft Manipulator

In our former research work, this book proposed a single module for foldable pneumatic soft manipulators, which is composed of one straight arm and one pneumatic actuator. The authors have proposed the kinematic model for a single module and have used the suspension method to conduct experiments. In space manipulation tasks, the properties of light weight and foldable volume can reduce the launching difficulties of the rockets, and the flexible movement enables the soft manipulator to conduct more complicated tasks. According to these considerations, this book present a design and development method appropriate for foldable soft manipulators. The great difficulties associated with the foldable soft manipulator lie in two aspects, the kinematic modelling and the ground-based experiment without considering gravity effects.

In this chapter, the design, modelling, and implementation of a foldable pneumatic soft manipulator are proposed. The soft manipulator comprises pneumatic actuators and foldable straight arms. The pneumatic actuator can achieve movement in all directions and the shape of the straight arm can be switched between the contracted and expanded states. The design parameters are important for the soft manipulator, especially for the pneumatic actuator. Based on this design, a kinematic model of the soft manipulator is established. The deformation and workspace of the soft manipulators are then numerically calculated and analyzed. Finally, a prototype of the foldable soft manipulator associated with the relevant air-floating devices is designed and developed. To validate the prediction accuracy of the kinematic model without considering the influence of gravity, the prototype of the soft manipulator is supported and floated on a marble platform using air-floating devices. By limiting the motion of the soft manipulator in a plane situation, the accuracy of the proposed model to predict the microgravity performance of the soft manipulator can be validated using the ground-based experiment. One kind of inverse kinematic modeling method, namely the IK-MLP-CR method, is proposed. It can classify the bending states and predict the whole manipulator shape by combining classification and regression based on multi-layer perceptron, which can greatly reduce the amount of

X. Zhang et al., *Design and Implementation of Soft Robotic Manipulators*,
https://doi.org/10.1007/978-981-95-1657-5_7

training data and training time. The performance of the proposed model is validated through the simulation and ground microgravity experiments.

7.1 Design of the Foldable Pneumatic Soft Manipulator

The foldable soft manipulator comprises three pneumatic soft modules. Each of the soft modules is composed of a pneumatic actuator and an inflatable straight arm, as shown in Fig. 7.1a. The soft manipulator can be transformed between the contracted and expanded states. The key component needed to achieve the switching of the state lies in the straight arm. The inflatable straight arm switches between the contracted state and the expanded state under the stimulus of the compressed air. When the straight arm is in the contracted state, the entire soft manipulator can be easily stored in the folding state, as shown in Fig. 7.1b. In the folding state, the straight arm is 120 mm long, while in the expanding state, it is 490 mm long. The length of the straight arm under the expanding state is four times longer than the folding state.

The design of the pneumatic actuator and the inflatable straight arm is shown in detail in Fig. 7.2. The pneumatic actuator comprises three inflatable bellows that have been linked through a set of rigid plates. The bellows extend when the pressured gas is inflated. The spatial movement of the actuator is produced based on the coupling effect of the extension of the three inflatable bellows and the constraints of the rigid plates, as shown in Fig. 7.2a. The inflatable arm comprises two layers,

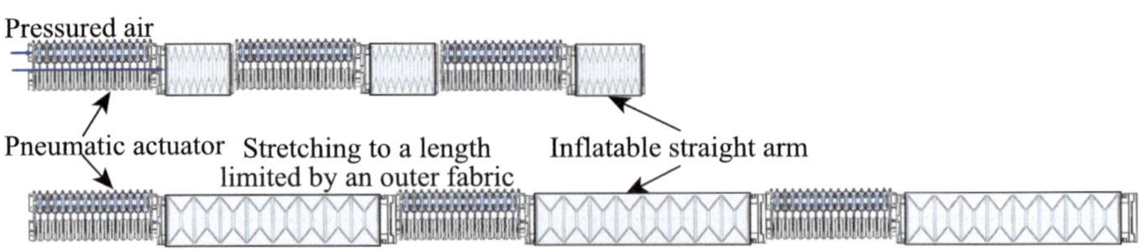

(a) Soft manipulator in the contracted and expanded states

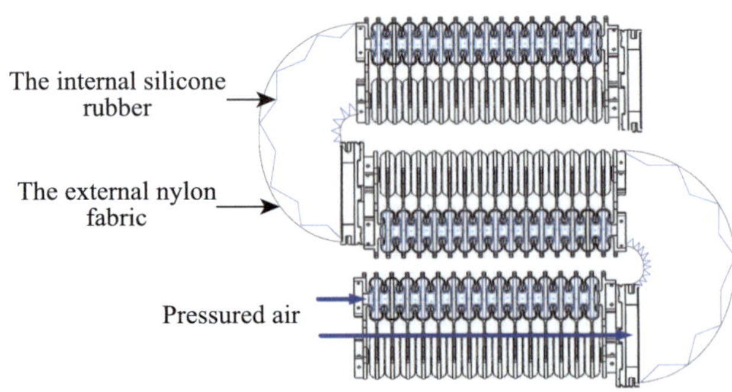

(b) Soft manipulator in a folding state

Fig. 7.1 Design of the foldable pneumatic soft manipulator

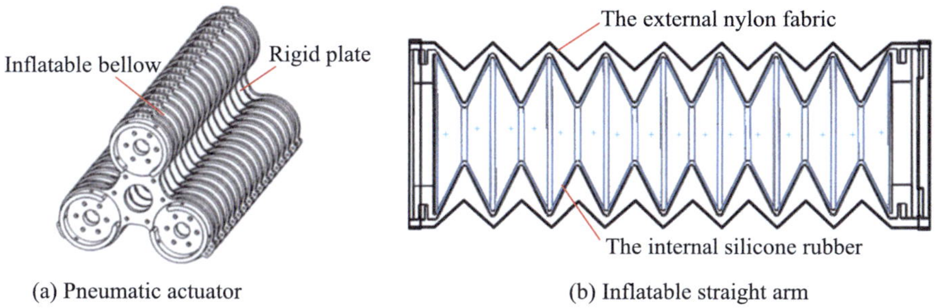

Fig. 7.2 Design of the pneumatic actuator and the inflatable straight arm

namely, internal and external, as shown in Fig. 7.2b. The internal layer is made by the soft silicone rubber and is responsible for the sealing function of the inflatable arm. The external layer is made of nylon fabric and is responsible for the structure enhancement, because the length of the inflatable straight arm is limited strictly by the external nylon fabric.

7.2 Kinematic Modelling of the Soft Manipulator

7.2.1 Kinematic Model

The definition of the inflatable bellow, the straight arm and the mathematical model of the basic inflatable unit are shown in Fig. 7.3.

In Fig. 7.3, the distance of two bellow centres is marked a, the original length of the bellow slice is L_0, and the elastic stiffness of the bellows is k. The bellows used in this chapter are all the same, so the k and L_0 involved are the same. It is assumed that the length increases of the three bellows are ΔL_1, ΔL_2, ΔL_3 when their input air pressures are P_1, P_2, P_3, respectively. Thus,

$$\begin{cases} P_1 S = k\Delta L_1 \\ P_2 S = k\Delta L_2 \\ P_3 S = k\Delta L_3 \end{cases} \tag{7.1}$$

where S is the cross-sectional area of two sides of the bellow. Then, the lengths of the bellows can be obtained as follows:

$$\begin{cases} L_1 = L_0 + \Delta L_1 = L_0 + P_1 S/k \\ L_2 = L_0 + \Delta L_2 = L_0 + P_2 S/k \\ L_3 = L_0 + \Delta L_3 = L_0 + P_3 S/k \end{cases} \tag{7.2}$$

The mechanical effect of the constraints of the rigid plates is a result of the cross-sectional centres of the bellows keeping an equilateral triangle and having the same

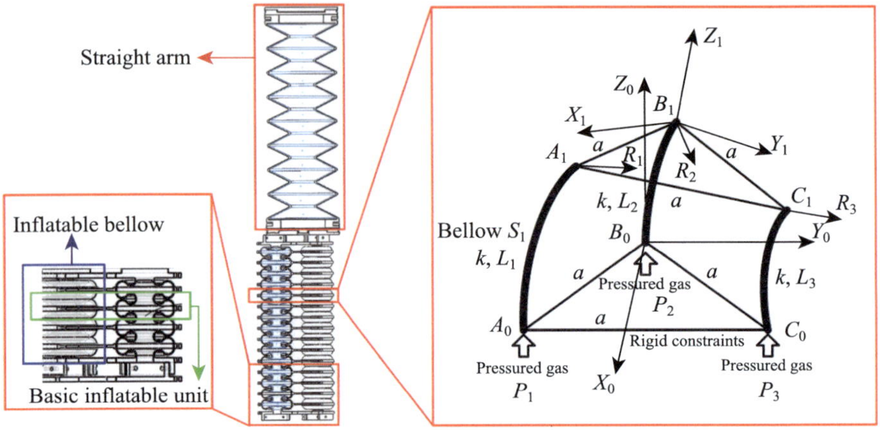

Fig. 7.3 Basic inflatable unit and inflatable bellow of the pneumatic actuator

normal direction. In Fig. 7.3, A_0, B_0, C_0 are the cross-sectional centres of three bellows. Using B_0 as the origin point, the body-fixed coordinate $X_0Y_0Z_0$ is defined. B_0Y_0 is parallel to A_0C_0. The applied forces of the rigid plates to the bellows are described as R_1, R_2, R_3, which are all in the cross section. The generated deflections are described as:

$$v_1 = \frac{R_1 L_1^3}{3EI} = \frac{R_1 L_1^2 S'}{3kI}, \quad v_2 = \frac{R_2 L_2^3}{3EI} = \frac{R_2 L_2^2 S'}{3kI}, \quad v_3 = \frac{R_3 L_3^3}{3EI} = \frac{R_3 L_3^2 S'}{3kI} \tag{7.3}$$

where, E is the elastic modulus of the bellows; I is the inertia moment of the cross section and EI is the bending stiffness; S' is the minimum cross-sectional area of the bellows shell, which is the effective area to afford the tension load. The slopes of the bellows generated by the forces R_1, R_2, R_3 are described as:

$$\begin{cases} \theta_1 = \dfrac{R_1 L_1^2}{2EI} = d\dfrac{R_1 L_1 S'}{2kI} \\ \theta_2 = \dfrac{R_2 L_2^2}{2EI} = \dfrac{R_2 L_2 S'}{2kI} \\ \theta_3 = \dfrac{R_3 L_3^2}{2EI} = \dfrac{R_3 L_3 S'}{2kI} \end{cases} \tag{7.4}$$

The cross sections of the bellows have the same normal direction, implying that the projections of three bellows in the X_0-Y_0 plane are parallel. Thus:

$$\{\theta_1 = \theta_2 = \theta_3\} \Rightarrow \{R_1 L_1 = R_2 L_2 = R_3 L_3\} \tag{7.5}$$

The projection angle of the bellows relative to the B_0Y_0 axis is given as α. When the bellows are inflated, the coordinates of A_1, B_1, C_1 are given by:

$$\begin{cases} A_1\colon \left[v_1 \cos\alpha + \dfrac{\sqrt{3}}{2} a \quad v_1 \sin\alpha - \dfrac{a}{2} \quad L_1 \right] \\ B_1\colon [v_2 \cos\alpha \quad v_2 \sin\alpha \quad L_2] \\ C_1\colon \left[v_3 \cos\alpha + \dfrac{\sqrt{3}}{2} a \quad v_3 \sin\alpha + \dfrac{a}{2} \quad L_3 \right] \end{cases} \tag{7.6}$$

According to the constraints $|A_1B_1| = |B_1C_1| = |C_1A_1| = a$, we have:

$$M_k^2 (L_1 - L_3)^2 W^2 - 2aM_k (L_1 - L_3) \sin\alpha W + (L_1 - L_3)^2 = 0 \tag{7.7}$$

$$M_k^2 (L_1 - L_2)^2 W^2 + aM_k (L_1 - L_2) \left(\sqrt{3}\cos\alpha - \sin\alpha\right) W + (L_1 - L_2)^2 = 0 \tag{7.8}$$

$$\tan\alpha = \frac{\sqrt{3}(L_1 - L_3)}{2L_2 - L_3 - L_1} \tag{7.9}$$

where

$$\begin{aligned} W &= R_1L_1 = R_2L_2 = R_3L_3 \geqslant 0 \\ M_k &= S'/(3kI) > 0 \end{aligned} \tag{7.10}$$

The variable W can be calculated by dividing into three cases:

(1) When $L_1 \neq L_3$, W can be calculated by Eq. (7.7).
(2) When $\Delta L_1 = \Delta L_3 \neq \Delta L_2$, W can be calculated by Eq. (7.8).
(3) When $\Delta L_1 = \Delta L_2 = \Delta L_3$, there is $W = 0$.

Once W is calculated, R_1, R_2, R_3, α can be calculated by Eqs. (7.9) and (7.10). The deflections v_1, v_2, v_3 can be calculated by Eq. (7.3) and the slopes θ_1, θ_2, θ_3 can be gotten by Eq. (7.4), it is already known that $\theta_1 = \theta_2 = \theta_3 = \theta$. Finally the coordinates of A_1, B_1, C_1 can be calculated by Eq. (7.6).

The pneumatic soft actuator is made up of a series of small bellow sections, as shown in Fig. 7.1b. According to the constant curvature assumption [1], all the small sections have the same curvatures. It is assumed that a vector in the second body-fixed coordinate $X_1Y_1Z_1$ is $\boldsymbol{r}_1 = [x_1 \quad y_1 \quad z_1]^{\mathrm{T}}$. When it is transferred to the first body-fixed coordinate $X_0Y_0Z_0$, it is defined as $\boldsymbol{r}_{10} = [x_{10} \quad y_{10} \quad z_{10}]^{\mathrm{T}}$; hence, we have:

$$\begin{aligned} \boldsymbol{r}_{10} &= \boldsymbol{H}^{\mathrm{T}}\boldsymbol{r}_1 + \overrightarrow{B_0B_1} \\ \boldsymbol{H} &= \boldsymbol{M}_3(\alpha)\,\boldsymbol{M}_2(\theta)\,\boldsymbol{M}_3(-\alpha) \end{aligned} \tag{7.11}$$

where, $\boldsymbol{H}$ is the transferring matrix between the two coordinates $X_0Y_0Z_0$ and $X_1Y_1Z_1$; $\overrightarrow{B_0B_1}$ is the vector from the origin point of $X_0Y_0Z_0$ to the origin point of $X_1Y_1Z_1$, there is:

$$\overrightarrow{B_0B_1} = [v_2 \cos\alpha \quad v_2 \sin\alpha \quad L_2]^{\mathrm{T}} \tag{7.12}$$

$\boldsymbol{M}_3(\alpha)$, $\boldsymbol{M}_2(\theta)$ and $\boldsymbol{M}_3(-\alpha)$ are the coordinate transferring matrixes, there is:

$$\boldsymbol{M}_3(\alpha) = \begin{bmatrix} \cos\alpha & -\sin\alpha & 0 \\ \sin\alpha & \cos\alpha & 0 \\ 0 & 0 & 1 \end{bmatrix}$$

$$\boldsymbol{M}_2(\theta) = \begin{bmatrix} \cos\theta & 0 & \sin\theta \\ 0 & 1 & 0 \\ -\sin\theta & 0 & \cos\theta \end{bmatrix}$$

$$\boldsymbol{M}_3(-\alpha) = \begin{bmatrix} \cos(-\alpha) & -\sin(-\alpha) & 0 \\ \sin(-\alpha) & \cos(-\alpha) & 0 \\ 0 & 0 & 1 \end{bmatrix} \tag{7.13}$$

It is known that the soft actuator has n sections in total. It was assumed that a vector in the body-fixed coordinate $X_{i-1}Y_{i-1}Z_{i-1}$ is also $\boldsymbol{r}_1 = [x_1 \quad y_1 \quad z_1]^{\mathrm{T}}$. When it is transferred to the first body-fixed coordinate $X_0Y_0Z_0$, it is defined as $\boldsymbol{r}_{i0} = [x_{i0} \quad y_{i0} \quad z_{i0}]^{\mathrm{T}}$, then we have:

$$\boldsymbol{r}_{i0} = \left(\boldsymbol{H}^{\mathrm{T}}\right)^{i-1}\boldsymbol{r}_1 + \left[\left(\boldsymbol{H}^{\mathrm{T}}\right)^{i-2} + \left(\boldsymbol{H}^{\mathrm{T}}\right)^{i-2} + \cdots + \left(\boldsymbol{H}^{\mathrm{T}}\right)^{2} + \left(\boldsymbol{H}^{\mathrm{T}}\right)^{0}\right] \cdot \overrightarrow{B_0B_1} \tag{7.14}$$

where, $2 \leqslant i \leqslant n$, and n indicates the total amount of bending units.

When the inflatable straight arm has been fastened onto the pneumatic actuator end, the elongation of this straight arm is tangent to the pneumatic actuator end. The straight arm, in which the stiffness subject to specific air pressures may reach a higher level in contrast to the pneumatic actuator, is reduced to a straight line during the kinematic modelling procedure. The free top coordinate in straight arm $\boldsymbol{r}_{\mathrm{TS}}$ is expressed as:

$$\boldsymbol{r}_{\mathrm{TS}} = \left(\boldsymbol{H}^{\mathrm{T}}\right)^{n}\begin{bmatrix} 0 \\ 0 \\ L_{\mathrm{S}} \end{bmatrix} + \left[\left(\boldsymbol{H}^{\mathrm{T}}\right)^{n-1} + \left(\boldsymbol{H}^{\mathrm{T}}\right)^{N-2} + \cdots + \left(\boldsymbol{H}^{\mathrm{T}}\right)^{1} + \left(\boldsymbol{H}^{\mathrm{T}}\right)^{0}\right] \cdot \overrightarrow{B_0B_1} \tag{7.15}$$

where, L_{S} is the length of the straight arm. Using Eqs. (7.14) and (7.15), the kinematic model [2] of the pneumatic actuator can be obtained. Then the kinematic model of the whole soft manipulator can be obtained by using coordinate system transformation.

7.2.2 *Numerical Calculation and Analysis*

7.2.2.1 Model Parameters

In the numerical calculation, three pneumatic actuators and two straight arms are chosen and connected to form the foldable soft manipulator, as shown in Fig. 7.4. To make the following analysis clear to understand, the bellows of the pneumatic actuator and the straight arm are numbered using specific symbols, as shown in Fig. 7.4.

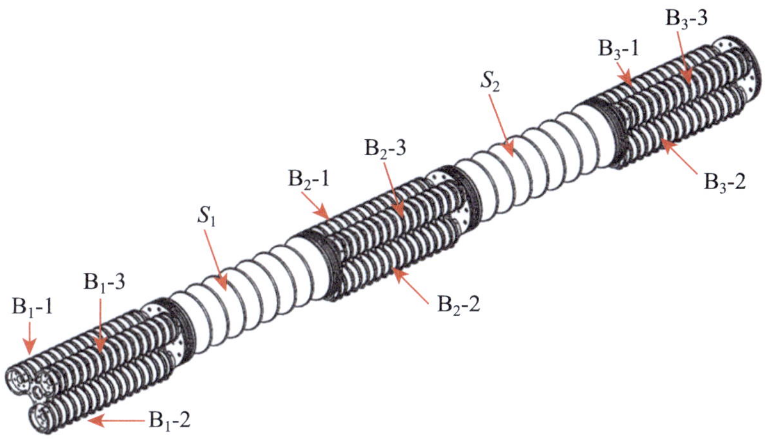

Fig. 7.4 Numbering role of the soft manipulator

Table 7.1 Model parameters of the soft manipulator

Parameter	L_0	a	n	$S_{1\text{empty}}$	$S_{2\text{empty}}$
Value	0.02 m	0.075 m	16	0.1 m	0.1 m
Parameter	d_{out}	d_{in}	T	$S_{1\text{full}}$	$S_{2\text{full}}$
Value	0.05 m	0.0488 m	0.02 m	0.490 m	0.485 m

In Fig. 7.4, it is observed that the numbers of the pneumatic actuator at the bottom part are B_1-1, B_1-2, and B_1-3. The numbers of the middle and top parts are B_2-1, B_2-2, and B_2-3 and B_3-1, B_3-2, and B_3-3, respectively. The numbers of the two straight arms are S_1 and S_2. This type of numbering role will be helpful in the following analysis. The model parameters of the soft manipulator are listed in Table 7.1.

In Table 7.1, the indications of the parameters L_0, a, and n can be found in Sect. 7.2.1. In addition, $S_{1\text{empty}}$ and $S_{2\text{empty}}$ are the lengths of the straight arms in the empty state; $S_{1\text{full}}$ and $S_{2\text{full}}$ are the lengths of the first and the second straight arms in their expanded states; d_{out} and d_{in} are the outer and the inner diameters of the bending unit; T is the height of the connection part between the pneumatic actuator and the straight arm. The elastic coefficient K of the pneumatic actuator is estimated to be equal to 23,000 N/m in this study.

7.2.2.2 Shape Deformation of the Soft Manipulator

Based on the model parameters in Table 7.1, the pneumatic actuator-related inputting pressure is set from 0.02 to 0.1 MPa, with intervals of 0.02 MPa, in the following numerical calculations. The straight arm is simplified as a straight line for calculation. Two different control strategies are designed to get different shape deformation of

Table 7.2 Numbers of the inflating bellows in the first control strategy

	Pneumatic actuator B_1	Pneumatic actuator B_2	Pneumatic actuator B_3
Bellow number	B_1-1	B_2-1	B_3-1
	B_1-2	B_2-2	B_3-2
	B_1-3	**B_2-3**	**B_3-3**

the soft manipulator. In the first control strategy, three bellows along the same central axis, whose numbers are marked in bold in Table 7.2, are inflated.

According to the control strategy in Table 7.2, the shape deformation of the soft manipulator is shown in Fig. 7.5.

In Fig. 7.5, it is observed that the shape deformation will increase as input pressure in the pneumatic actuators increases. The soft manipulator makes the simple bending movement because the three inflating bellows are along the same central axis. According to the results shown in Fig. 7.5b, the projection of the shape deformation in the X-Y plane is a straight line, while the projections in the X-Z and Y-Z planes are nearly identical. This indicates that no disturbance occurs in the numerical results of the kinematic model, which are consistent with the physical law of the soft manipulator. In the second control strategy, another three bellows are inflated, and

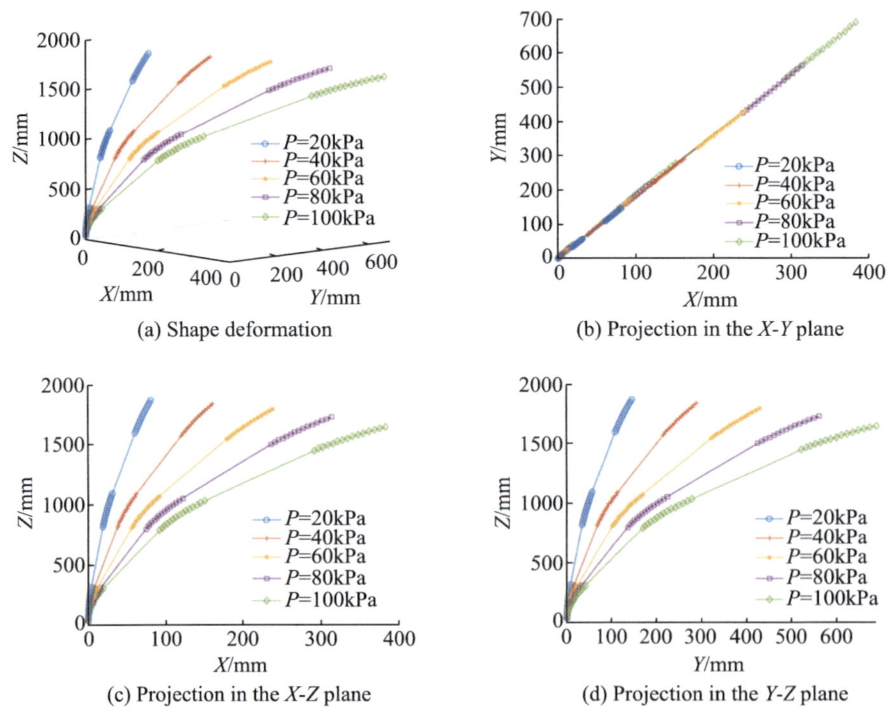

Fig. 7.5 Shape deformation in the first control strategy

Table 7.3 Numbers of the inflating bellows in the second control strategy

	Bending joint B_1	Bending joint B_2	Bending joint B_3
Bellow number	**B_1-1**	B_2-1	B_3-1
	B_1-2	**B_2-2**	B_3-2
	B_1-3	B_2-3	**B_3-3**

the bellow numbers are marked in bold, as listed in Table 7.3. The input pressure of the bellows is set from 0.02 to 0.1 MPa at 0.02 MPa intervals.

According to the control strategy in Table 7.3, the shape deformation of the soft manipulator is shown in Fig. 7.6.

In Fig. 7.6, it is observed that the shape deformation of the soft manipulator in the second strategy exhibits a complicated situation. The reason is attributed to the fact that the three inflating bellows are located along a spiral condition. Figures 7.6b–d show the projections of the deformation in three different planes. The different shapes of these projections indicate that the deformation of the soft manipulator occurs in three-dimensional space and thus appears as a spiral condition. The deformation

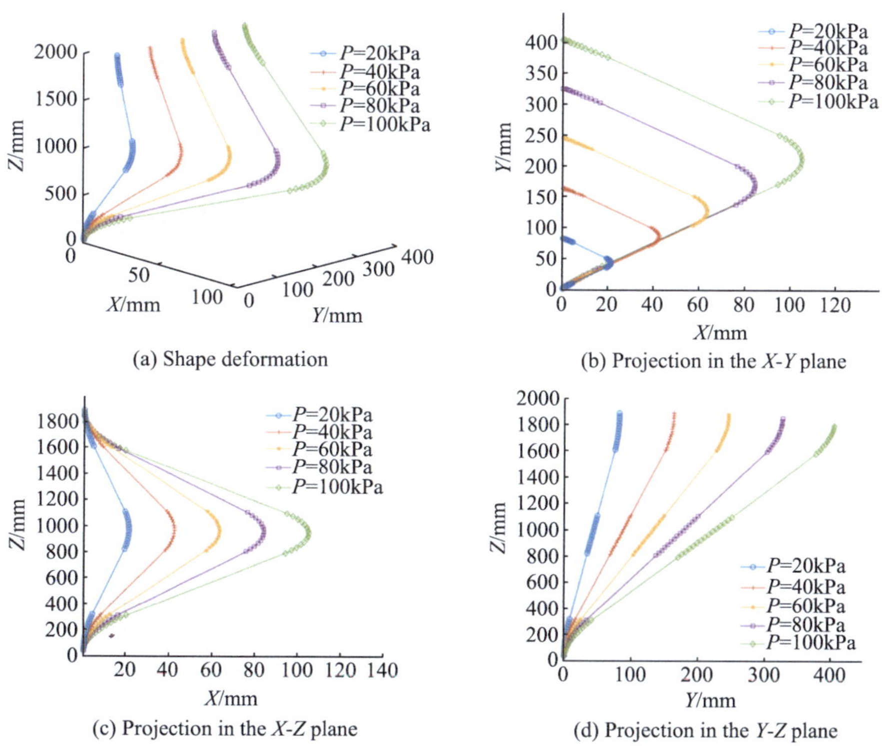

Fig. 7.6 Shape deformation in the second control strategy

magnitude of the soft manipulator increases as the input pressure of the bellows increases. These numerical results of the kinematic model are also consistent with our expectations.

7.2.2.3 Workspace of the Soft Manipulator

The special advantage of the foldable pneumatic soft manipulator is that it can be kept in the contracted state and can switch to the expanded working state. When it is in the contracted state, the soft manipulator can reduce storage volume; When it switches to the expanded state, the soft manipulator can conduct the operating tasks in a wide workspace. In the following, the workspaces of the soft manipulator in the contraction state and the expanded state are calculated and analyzed, as shown in Fig. 7.7.

In Fig. 7.7, the pneumatic actuators are inflated using air pressure in the range of 0.01–0.1 MPa at random, with an interval of 0.01 MPa. The arc surface labelled by the blue star represents the workspace in the expanded state, while the arc surface labelled by the red star represents the workspace in the contraction state. Fig. 7.7a compares the three-dimensional view of two kinds of workspaces; Fig. 7.7b and c show the projections of the workspace in three different planes. From the above comparisons, it is clear that the soft manipulator has a much wider workspace when it is in the expanded state.

7.3 Implementation and Experimental Validation

7.3.1 Experiment Design

The experimental system is mainly composed of a prototype of a soft manipulator, a pneumatic control system, two air-floating devices, and a VICON positioning device. The prototype of the soft manipulator, which has a total length of 2.3 m, comprises three pneumatic actuators as well as two straight arms. The pneumatic actuator is fabricated by using SLS print technology. The straight arm is constructed based on the combination of two levels, including an inner sealing level and an outer fabric level. The prototype of the soft manipulator has a total length of 2.3 m, a total weight of 5 kg and a diameter of 20 cm, while the Canadarm has a total length of 17.6 m, a total weight of 1.8 t and a diameter of 35 cm on average [3]. It can be seen that the soft manipulator is far lighter at the same length. The foldability of the straight arm is realized by the soft property of the structure and materials when it is not inflated, as shown in Fig. 7.8a. To make the soft manipulator move freely in the plane situation, it is supported with the use of two air-floating devices, as shown in Fig. 7.8b. The principle of the pneumatic control system is shown in Fig. 7.8c. It can be seen that there are a total of 11 gas channels for the pneumatic control of the soft

Fig. 7.7 Workspace of the end point of the soft manipulator

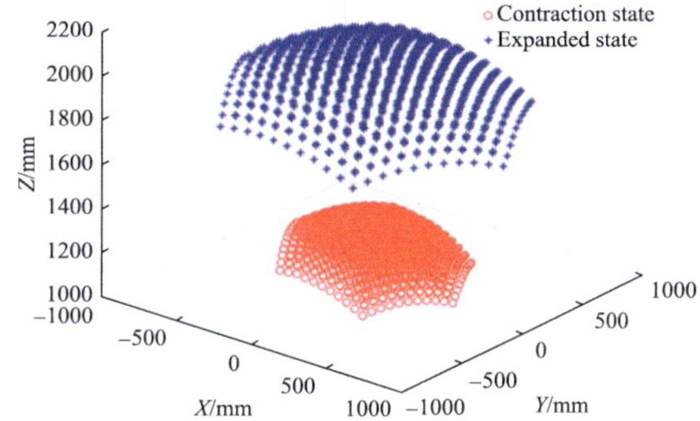

(a) Workspace distribution

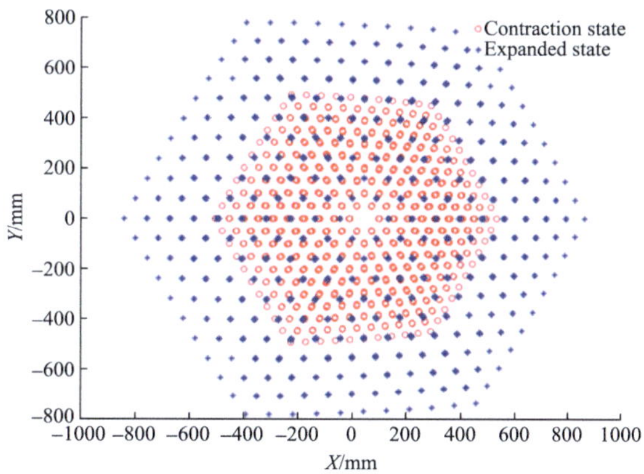

(b) Projection of the workspace in the X-Y plane

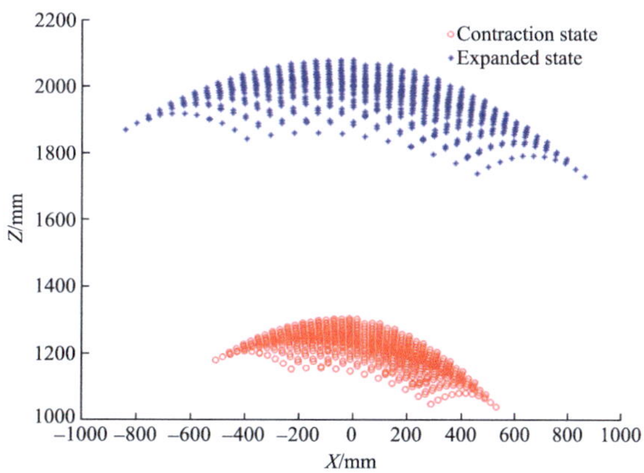

(c) Projection of the workspace in the X-Z plane

Fig. 7.8 Experimental system of the soft manipulator

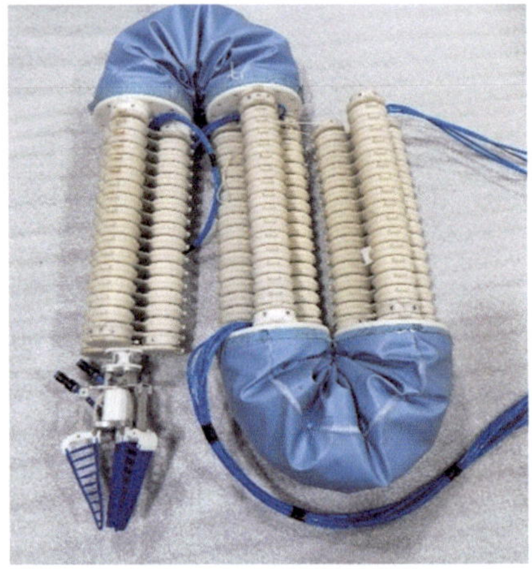

(a) Soft manipulator in contracted states

(b) Experimental prototype

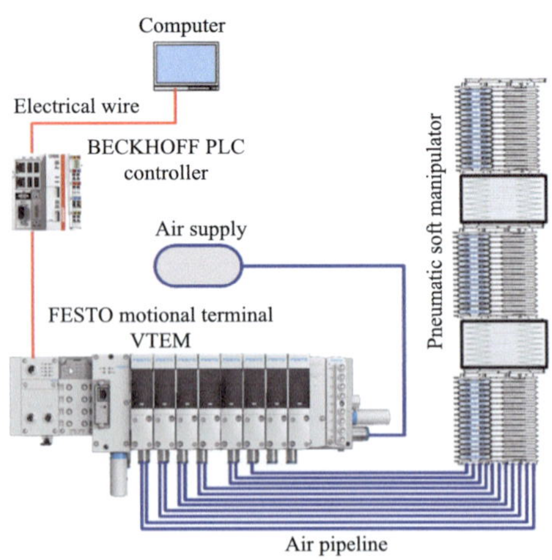

(c) Pneumatic control system

manipulator, including nine gas channels for three pneumatic actuators, and two gas channels for two straight arms.

In Fig. 7.8, it is seen that the prototype of the soft manipulator is floating on the marble platform using the air-floating devices, which can move freely with negligible friction force in the plane situation. Thus, the soft manipulator can have three DOFs in this experimental environment, including two DOFs in the plane and one rotation freedom along the normal direction of the plane. In the experiment system, the marble platform is extremely flat, with a height difference less than 2.8 μm. The air-floating device includes a spherical gas tank, a support platform, three jet valves, several solenoid valves and pressure gauges, as shown in Fig. 7.9a. The envelope size of the device is 26 cm × 26 cm × 60 cm and it floats on the marble platform by releasing the pressured air from the jet valves. To obtain the shape deformation of the soft manipulator, the VICON positioning device is used, as shown in Fig. 7.9b. It is produced by VICON Motion Systems Ltd., UK. A series of markers are trapped on the experimental object to obtain its position coordinates. Its positioning precision can reach 1 mm.

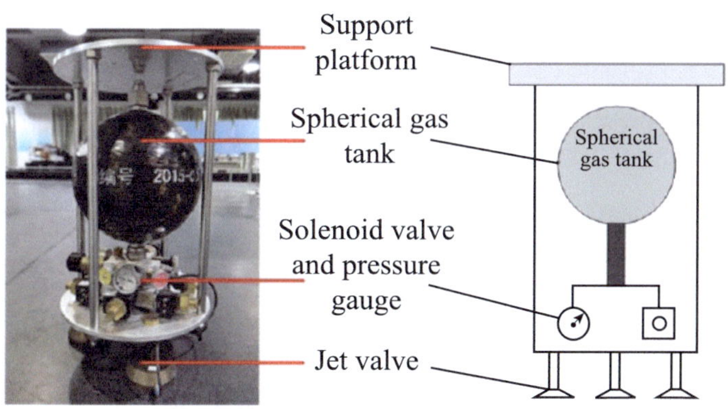

(a) Air-floating device

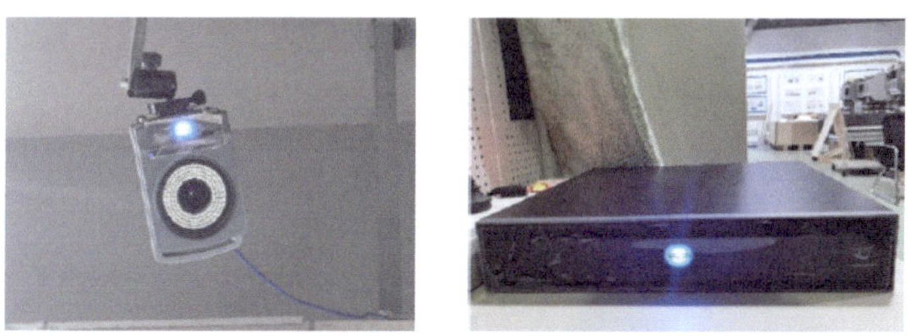

(b) VICON positioning device

Fig. 7.9 Experiment auxiliary devices

7.3.2 Experimental Validation of the Soft Manipulator

The performance of the soft manipulator is verified based on the above experimental system. The tests are conducted in two control strategies in which the numbers of the inflating bellows are different. The shape deformation of the soft manipulator is documented for a comparison with the calculated results of the kinematic model. The experiment is performed on the marble platform to simulate the free motion in the plane condition. The numerical results of the kinematic model are also converted to the horizontal plane to compare with the experimental results.

7.3.2.1 The First Control Strategy: The Pendulum Movement

In this control strategy, the numbers of the inflating bellows are marked in bold and listed in Table 7.4. It is noted that the bellows B_1-1, B_2-1, and B_3-1 are along the same central axis, and the bellows B_1-2, B_2-2, and B_3-2 are also along the same central axis. The coupling effect of the inflating of these six bellows with the same air pressure is to make the soft manipulator perform the pendulum movement in the plane state, thus the influence of the gravity can be ignored.

According to the control strategy in Table 7.4, the inflating pressures of the bellows are set from 0.02 MPa to 0.1 MPa at 0.02 MPa intervals. The shape deformation and comparison between the model predictions and the experimental results are shown in Fig. 7.10.

Figure 7.10a shows the experimental scene of the soft manipulator in the first control strategy and Fig. 7.10b shows the shape deformation comparison between the model predictions and the experimental results. In Fig. 7.10b, the experimental results are depicted using lines marked with symbols, while the numerical calculation results are depicted using curves without markers. It is observed that the soft manipulator makes pendulum movements in this control strategy, which is consistent with the model prediction. As the inflating pressure increases, the bending amplitude increases. At this point, the model predictions also agree with the experimental results. The ending points distribute from 1800 to 2000 mm in the Y axis and from −180 to −870 mm in the X axis. From the results shown in Fig. 7.10b, it is observed that the difference between the model prediction and the experiment results still exists. The main reason lies in the fact that the markers installed on the soft manipulator are impossible to set exactly alongside its central axis. In terms of bending

Table 7.4 Numbers of the inflating bellows in the first control strategy

	Bending joint B_1	Bending joint B_2	Bending joint B_3
Bellow number	**B_1-1**	**B_2-1**	**B_3-1**
	B_1-2	**B_2-2**	**B_3-2**
	B_1-3	B_2-3	B_3-3

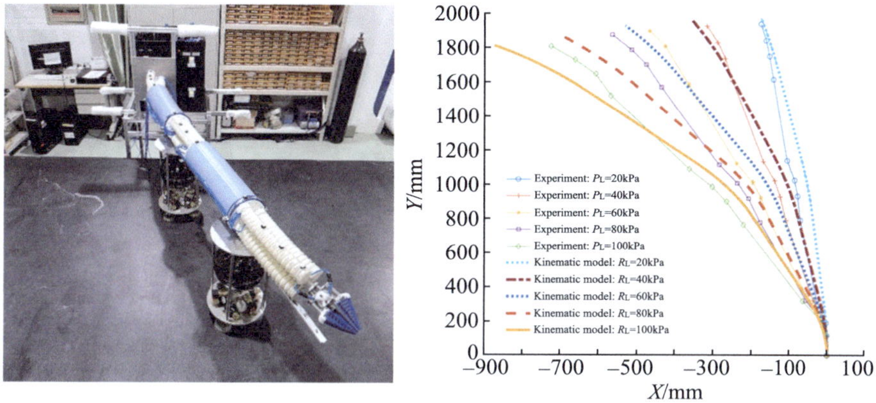

(a) Shape deformation (b) Deformation comparison

Fig. 7.10 Shape deformation and comparison between the model predictions and the experimental results in the first control strategy

patterns and the bending trends, the numerical results of the kinematic model agree well with the experimental results.

7.3.2.2 The Second Control Strategy: The "S" Shape Bending

In this control strategy, the numbers of the inflating bellows are marked in bold and listed in Table 7.5. It is noted that the bellow B_2-3 is in the parallel-axial symmetry mode with B_2-1 and B_2-2. The actuation effect of B_2-3 is exactly opposite to the coupling effect of B_2-1 and B_2-2. All these bellows chosen in Table 7.5 are inflated with the same air pressure each time.

According to the control strategy in Table 7.5, the inflating pressures of the bellows are determined in the scope of 0.02–0.1 MPa, with an interval of 0.02 MPa. The shape deformation and comparison between the model predictions and the experimental results are shown in Fig. 7.11.

Figure 7.11a shows the experimental scene of the soft manipulator in the second control strategy and Fig. 7.11b shows the shape deformation comparison between the model predictions and the experimental results. In Fig. 7.11b, the experimental results are depicted using lines marked with symbols, while the numerical calcula-

Table 7.5 Numbers of the inflating bellows in the second control strategy

	Bending joint B_1	Bending joint B_2	Bending joint B_3
Bellow number	B_1-1	**B_2-1**	B_3-1
	B_1-2	**B_2-2**	B_3-2
	B_1-3	B_2-3	**B_3-3**

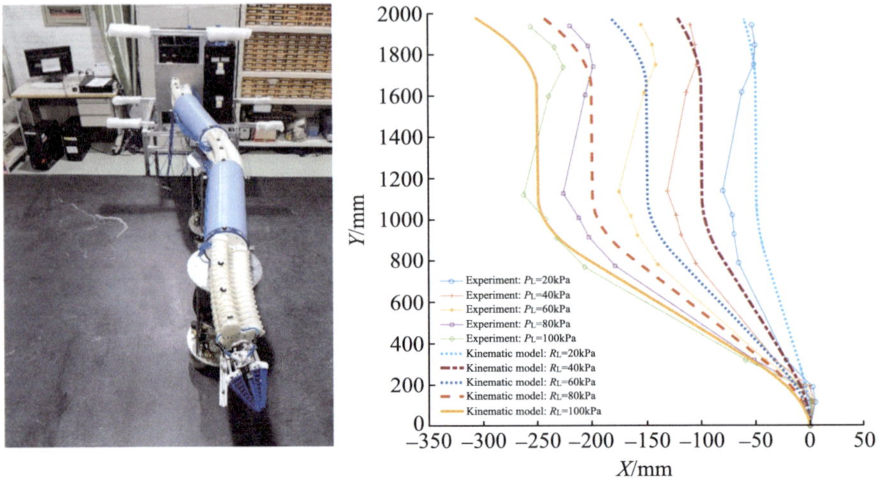

(a) Shape deformation (b) Deformation comparison

Fig. 7.11 Shape deformation and comparison between the model predictions and the experimental results in the second control strategy

tion results are depicted using curves without markers. It is observed that the soft manipulator deforms to a reversed "S" shape in this control strategy. Additionally, the bending amplitude increases with the increasing inflating pressure. Comparing the deformation patterns and the bending trends, the model prediction and the experimental results agree well with each other. This verifies the correctness of the kinematic model of the soft manipulator. The difference between the model prediction and the experiment still exists, especially at the end points and the middle points. The main reason is also that the markers installed on the soft manipulator are impossible to set exactly alongside its central axis. And the existing friction inside the soft manipulator structures may lead to discrepancies between the numerical and experimental results. Moreover, the difference at the end point is magnified by the length of the soft manipulator, while the difference at the middle point is easy to see because it locates the turning part of the reversed "S" shape.

7.4 Inverse Kinematics Modeling for Overall Shape Prediction of Soft Manipulators

The inverse kinematics modelling of soft manipulators through traditional methods is subject to truncation errors, complex derivations and inexpressibility equations, owing to geometric singularities. Furthermore, facing the future control requirement for multiple major feature points on manipulators, most neural network approaches that solve the drive space only by the inverse solution of the end task space, cannot obtain the inverse kinematics solutions of all feature points, due to their including redundancy (the dimension n of the drive space of feature points is larger than that of the end task space). Therefore, this chapter proposes a method combining classification and regression based on multi-layer perceptron for inverse kinematic modelling, namely the IK-MLP-CR method, with the bending states classified and the whole manipulator shape predicted, greatly reducing the amount of training data and training time. Finally, the performance evaluation of the proposed model is validated through the simulation and ground microgravity experiments.

For kinematic modeling, highly accurate methods such as the FEM [4] and the geometrically exact method [5] are often accompanied by over-computation to ensure accuracy, which makes it difficult to provide real-time feedback on position and orientation for sub-sequent control. An efficient model can greatly save computation overhead for fine-grained simulations [6, 7]. In contrast, the piecewise constant curvature (PCC) method [2], benefiting from the simplicity and efficiency of having only three configuration variables, e.g. in pneumatic [8] and tendon [9] actuators, sacrifices accuracy but has gained wide acceptance in terms of practicality. These theoretical analysis methods discussed are common for modelling forward kinematics, but when involving inverse kinematics (IK), inverse kinematic direct derivation becomes very complex and does not lead to the closed-form IK solution [10]. Therefore, in the vast majority of cases where this IK solution is not available, numerical methods, such as the Jacobian-based matrix method [11] and iterative methods [12], as well as the modal method [13], are required. Inevitably, as the target pose approaches the singular shape under pure elongation, the Jacobian matrix takes longer and is hard to converge and the modal equation is not always available.

Considering the problems with traditional methods, some simple neural network (NN) approaches [14, 15] are applied to inverse kinematic modeling. Nevertheless, these approaches are limited to considering the endpoint location. When manipulators with multiple segments are used for shape measurement, it is necessary to consider not only the endpoint position of the soft manipulator, but also the multiple feature points of the overall shape of multi-segment manipulators. Otherwise, it will result in redundancy (the dimension n of the drive space is larger than that of the task space), a problem that has been solved in conventional methods. Therefore, a new neural network inverse kinematic model for multi-segment manipulators is required, and presented here with the structural accuracy of the PCC method and generalization abilities of the NN (including classification and regression). It not only circumvents the kinematic singularities in PCC models, but also predicts multiple feature points for overall shape of multisegment manipulators.

7.4.1 Coordinate Frames Definition

The soft manipulator is divided into three actuators which are made of pneumatic bending segments and are connected by rigid restraining frames. The rigid restraining frames maintain overall spatial motion without bending or deformation during arm

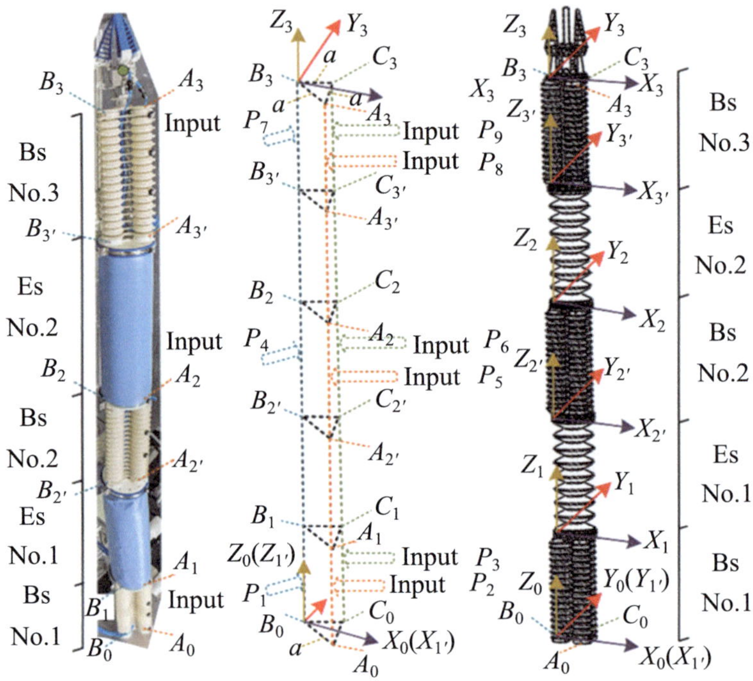

Fig. 7.12 Coordinate frames and input pressures of the bending segments (Bs) and extended segments (Es)

movement. Between two adjacent rigid restraining frames, the airbags of the bellows can stretch or bend. As shown in Fig. 7.12, bending segments (No. 1, No. 2 and No. 3 marked with purple) and two extended segments ((No. 1, No. 2 marked with dark yellow) can expressed in turn as $A_0B_0C_0$-$A_1B_1C_1$, $A_{2'}B_{2'}C_{2'}$-$A_2B_2C_2$, $A_{3'}B_{3'}C_{3'}$-$A_3B_3C_3$ and $A_1B_1C_1$-$A_{2'}B_{2'}C_{2'}$, $A_2B_2C_2$-$A_{3'}B_{3'}C_{3'}$, respectively in Fig. 7.12. Each bending segment is an actuator and consists of three bellows symmetrically with a spacing a between them. Each bellow is controlled by a separate pneumatic pressure with airbags, and the soft manipulator has a total of nine independent pneumatic control inputs, which are used to control different combinations of pneumatic pressure to realize complex spatial movements of the arm. For consistency, the following actuators are described by segments because they are driven by air pressure and can also be regarded as actuators.

In addition, a bending segment and an extending segment, abbreviated as Bs. and Es. respectively, form a basic unit of the whole manipulator. In Fig. 7.12, A_0, B_0 and C_0 are the centers of the fixed ends of the three airbags, and the three points form an equilateral triangle with side length a. With B_0 as the coordinate origin, the X_0-axis is perpendicular to A_0C_0, the Y_0-axis is perpendicular to the X_0-axis to the right, and the Z_0-axis is in the direction of the length of the arm, constituting a right-handed right-angled coordinate system. The mechanical effect of the rigid restraining frame

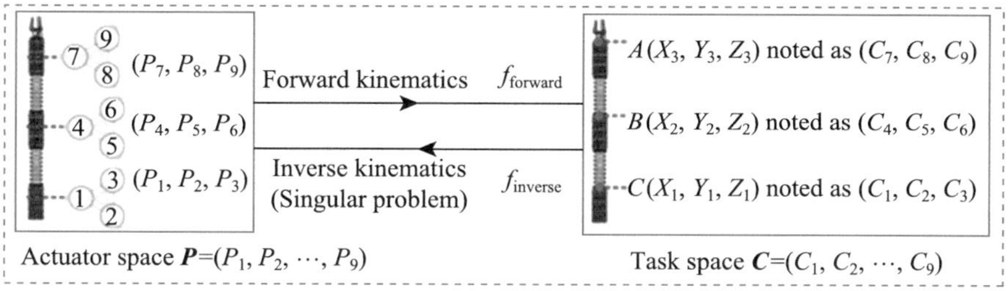

Fig. 7.13 The forward and inverse kinematic layout of the soft manipulator

$A_3B_3C_3$ at the upper end is twofold: ① to make the three points A_3, B_3 and C_3 at the upper end of the three airbags still form an equilateral triangle with side length a; ② to make the three airbags have the same spatial pointing in the normal direction.

In Fig. 7.13, the positions of the feature points $A(X_1, Y_1, Z_1)$, $B(X_2, Y_2, Z_2)$, and $C(X_3, Y_3, Z_3)$ at the end positions of the three bending segments No.1~No.3, noted as $(C_1, C_2, \cdots, C_9)$, are respectively located at the center of the triangles $A_1B_1C_1$, $A_2B_2C_2$, $A_3B_3C_3$ from Fig. 7.12. These three feature points A, B, and C together describe the rough shape of the soft manipulator.

Meanwhile, Fig. 7.13 involves the forward and inverse kinematic layout of the positions of the end feature points A, B, C, characterizing the interrelationship between the actuator space $\boldsymbol{P} = (P_1, P_2, \ldots, P_9)$ and task space $\boldsymbol{C} = (C_1, C_2, \ldots, C_9)$. The forward kinematic expression $C = f_{\text{forward}}(P)$ and the inverse kinematics $P = f_{\text{inverse}}(C)$ among the feature points A, B, and C can be obtained as follows:

$$(C_1, C_2, \ldots, C_9) = f_{\text{forward}}(P_1, P_2, \ldots, P_9) \tag{7.16}$$

$$(P_1, P_2, \ldots, P_9) = f_{\text{inverse}}(C_1, C_2, \ldots, C_9) \tag{7.17}$$

where f_{forward} and f_{inverse} are the forward and inverse kinematic mapping functions, respectively. And $(P_1, P_2, \ldots, P_9)$ is the pressure sequence which can be noted as P, and $(C_1, C_2, \ldots, C_9)$ is the sequence of the position coordinates which can be noted as C.

Based on the PCC method, this section proposes a fitting model for Inverse Kinematics-Multilayer Perceptron combined with Classification and Regression, namely the IK-MLP-CR model, which offers a model-free and iterative way to predict the input pressures according to the given position of the bending segment. And the proposed method not only saves computation time but avoids the geometric singular problem caused by the traditional inverse kinematic calculation method. The main established process of the data preparation, model organization and model performance are shown in Fig. 7.14 and the detailed operations are expanded as follows.

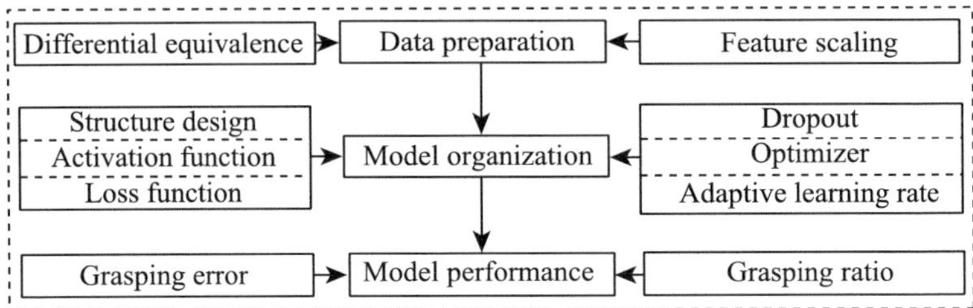

Fig. 7.14 The process of building the IK-MLP-CR model

7.4.2 Equivalent Simplification and Data Preparation

Equivalent simplification refers to the quantification of the difference in input air pressures. It not only can help achieve better training results with fewer data dimensions, which can effectively prevent overfitting of the neural network during training, but also reduce the dimensionality of the data and improve the training accuracy by simplifying the different equivalent inflatable combinations. The whole unilateral planar bending of the soft robotic arm is mainly determined by the value of the three-dimensional pressure. The pressure of one section of the bending segment can be simplified to be mainly determined by P_1, similarly, the air pressure of the second section and the third section of the bending joint can be determined by P_4 and P_7, respectively, and the combination of the three-section actuator input air pressures is 23^3=12,167 kinds of possibilities. It can be seen that the process of the equivalent simplification can lead to a significant reduction in the dimension of the training data and a significant reduction in the training time.

With the above technique, the air pressure input sequence (P_1, P_2, ..., P_9) can be reduced to a sequence of only 3 air pressures (P_1, P_4, P_7), which can be used to achieve in-plane bending motion to the right. Similarly, the sequence (P_2, P_5, P_8) facilitates the realization of in-plane leftward bending motion due to structural symmetry. Therefore, the sequence (P_2, P_3, P_5, P_6, P_8, P_9) can be simplified and expressed only by (P_2, P_5, P_8). And the simplified forward kinematic relationship between the feature positions A, B, and C (in Fig. 7.12) and pressures can be expressed as:

$$C = (C_1, C_2, \ldots, C_9) = \begin{cases} f\,(P_1, P_4, P_7), \text{ Right bending} \\ f\,(P_2, P_5, P_8), \text{ Left bending} \end{cases} \tag{7.18}$$

where (C_1, C_2, ..., C_9) represent the specific position sequence of feature points A, B, C (in Fig. 7.12), (C_1, C_2, ..., C_9) can be noted as C; (P_1, P_4, P_7) are the air pressures which result in the bending to the right; (P_2, P_5, P_8) are the air pressures which result in the bending to the left.

Referencing the kinematic model in Sect. 7.2 [16], 80 and 20% of the total data samples are selected as the training and test sets respectively in this chapter (it is important to have not only a large proportion of the training dataset, but also to ensure a certain number of test datasets to ensure good generalization abilities). Furthermore, feature scaling (data normalization) is required due to position coordinates and input pressures are not of the same order of magnitude. Among the samples, the values of C range from -2 to 2 m, while the values of P range from 20,000 Pa to 130,000Pa. The min-max normalization on the output variables P'_{ij} can be:

$$P'_{ij} = \frac{P_{ij} - P_{j\min}}{P_{j\max} - P_{j\min}} \tag{7.19}$$

where, P_{ij} represents the pressure for different training groups i (i =1, 2, ..., 12,167) and different dimensions j(j =1, 2, 3); $P_{j\max}$ and $P_{j\min}$ are the maximum and minimum air pressure values, respectively.

7.4.3 Model Organization

After obtaining the training and test sets based on the kinematic model, the IK-MLP-CR model is established as an inverse kinematic prediction algorithm, integrating classification and multilayer perceptron regression, which is trained in a fully supervised manner. The first classification model, according to the positive/negative abscissa C_1 of the first feature point A (in Fig. 7.12), classifies bending to the left (C_L) and bending to the right (C_R), as shown in Eq. (7.20). This right and left bend is noted as 1 and -1 in Eq. (7.21). The classified position-pressure sequence ($C \sim P$) is then passed into the regression neural network as afferent and efferent neurons, and the output of the mth neuron in the nth layer of this network, c^n_m, is given in Eq. (7.22). Figure 7.15 shows the structural design of the inverse kinematic prediction model. The input layer of the IK-MLP-CR network has 9 nodes corresponding to the end positions of the 3 bending segments; The hidden layer has 5 layers with [30, 60, 90, 60, 30] nodes and the Sigmoid activation function is to help add non-linearity to each node; The output layer has 3 nodes with the linear output activation function. To avoid overfitting problems, the dropout method is used so that the neurons in different layers are randomized through the probability P of deleting a neuron. The back-propagation algorithm [17] is implemented by the stochastic gradient descent (SGD) method [18] to achieve the structure and parameter determination of the inverse model. The IK-MLP-CR model uses an adaptive learning rate to obtain a fast convergence rate.

The determination of different hyperparameters in neural networks include: ① learning rate η; ② regularization parameter λ; ③ number of layers of the neural network n; ④ number of neurons in each hidden layer m; ⑤ number of rounds of learning Epoch; ⑥ size of minibatch; ⑦ coding method of the output neurons; ⑧ selection of the loss function; ⑨ method of weight initialization; ⑩ types of activation function; ⑪ size of data participating in the training model. Determination of these 11 types of hyperparameters requires a continuous and complex parameter-tuning process. It's crucial to monitor the training process and adjust the parameters based on the observed behaviour of the loss. At the same time, performing abla-

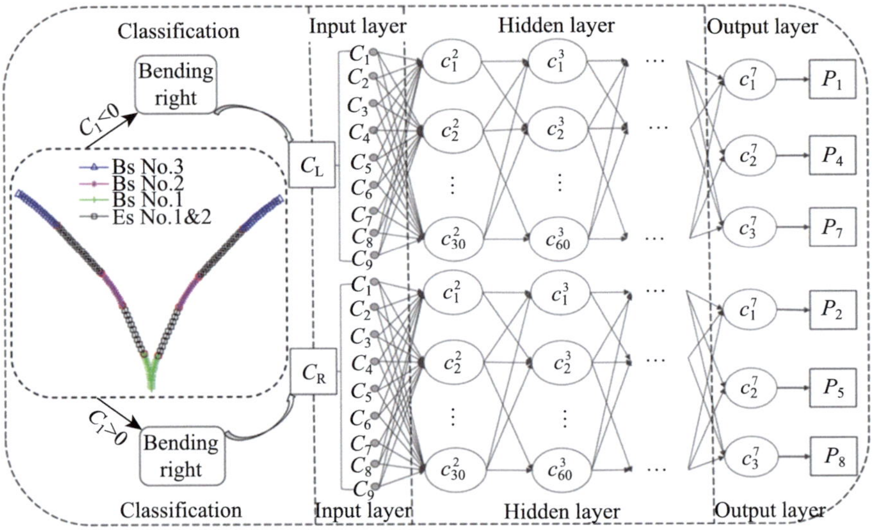

Fig. 7.15 The structure design of the IK-MLP-CR prediction model

Table 7.6 Details of hyperparameters of kinematic model

Hyperparameters	Values
Batch size	128
Learning rate	0.001
Epoch	1000
Weight decay	0.0001

tion experiments by systematically varying a single controlled variable is essential for determining optimal values tailored to specific models and tasks. After a lot of parameter tuning, hyperparameters in Table 7.6 such as the batch size, the initial learning rate, the number of iterations, and the weight decay can be set to 128, 0.001, 1000, and 0.0001, respectively.

$$P = \begin{cases} (P_1, P_4, P_7) = g(C_1, C_2, \ldots, C_9),\ C_1 > 0, \text{Rightward bending} \\ (P_2, P_5, P_8) = g(C_1, C_2, \ldots, C_9),\ C_1 < 0, \text{Leftward bending} \end{cases} \tag{7.20}$$

where $(C_1, C_2, \ldots, C_9)$ denotes the position sequence consisting of the horizontal (X-axis), longitudinal (Y-axis), and vertical (Z-axis) coordinates of the feature points A, B, C (in Fig. 7.12), which can be simply expressed as the position sequence C. And P including the air pressure sequences (P_1, P_4, P_7) and (P_2, P_5, P_8), which result in the rightward and leftward bending, respectively.

$$\mu_i = \begin{cases} 1, & C_1 > 0, \text{Rightward bending} \\ -1, & C_1 < 0, \text{Leftward bending} \end{cases} \tag{7.21}$$

where the positive and negative values of the horizontal coordinate C_1 corresponds to 1 and -1 of μ, respectively, implying a bend to the right or to the left.

$$c_m^n = \varphi(z_m^n) = \varphi\left(\sum_{r=1}^{k} w_{mr}^n c_r^{n-1} + b_{mr}^n\right) \tag{7.22}$$

where w_{mr}^n and b_{mr}^n respectively denote the weights and bias of the mth neuron in the nth layer connecting the ith neuron in the $(n-1)$th layer, and φ is the activation function. If n=2, the corresponding $c_r^1(r=1, 2, \ldots, 9)$ is the $C_i(i=1, 2, \ldots, 9)$ of the input layer; If n=3, the corresponding $c_r^2(r=1, 2, \ldots, 30)$ is the output of the ith neuron of the second layer; If n=2, 3, ..., 7, the corresponding $c_r^{n-1}(r=1, 2, \ldots, k)$ is the output of the ith neuron of the $(n-1)$th layer, and finally c_1^7, c_2^7, c_3^7 of the output layer are the outputs corresponding to the right-bending air pressure P_1, P_4, P_7 or left-bending air pressure P_2, P_5, P_8.

The mean square error (MSE) that is sensitive to outliers (squared operation) for the predicted pressure $\widehat{p}$ and true pressure p ensures that the error between these two values can be propagated backwards, making the NN simple and effective. Thus, the loss function can be expressed as:

$$\text{loss} = \frac{1}{n}\sum_{i=1}^{n}\left[(\widehat{p_i} - p_i)^2 + (E(\widehat{p}) - C)^2\right] \tag{7.23}$$

$$E(\widehat{p}) = \frac{1}{n}\sum_{i=1}^{n}\widehat{p_i} \tag{7.24}$$

where n is the number of samples, p_i and $\widehat{p_i}$ are the true and the predicted output pressure value respectively, $|E(\widehat{p}) - C|$ means the absolute value between the mean predicted value and inherent pressure.

Therefore, based on the PCC method, this chapter proposes a fitting model for inverse kinematics-multilayer perceptron combined with classification and regression, namely the IK-MLP-CR model, which offers a model-free and iterative way to predict the input pressures according to the given position of the bending segment. And the proposed method not only saves computation time but avoids the geometric singular problem caused by the traditional inverse kinematic calculation method.

7.4.4 *Model Performance and Discussions*

The performance evaluation of the proposed model is validated through the simulation and ground microgravity experiments. In the simulation, training and test data are provided based on the kinematic model indicated in Sect. 7.2. When the manipulator is trained to reach the predicted position C_{d1}, it is compared with the real position C_d and their minimum Euclidean distance, i.e. the grasping error e_l in Eq. (7.25). The samples that have an error of grasping within 7mm are defined as successful grasping. When aiming at the targeted objects, the successful grasping ratio can be Eq. (7.26).

$$e_l = \sqrt{(C_{dx} - C_{d1x})^2 + (C_{dy} - C_{d1y})^2} \quad (7.25)$$

where C_{dx}, C_{dy}, C_{d1x} and C_{d1y} are the x and y coordinates of the true reaching position C_d and the predicted reaching position C_{d1}, respectively.

$$\text{GR} = n_{\text{grasping}}/N \quad (7.26)$$

where n_{grasping} is the number of samples to reach the target.

Corresponding to the leftward and rightward bending, their performance curves of the loss function and accuracy on the 300-iterations and 1000-iterations training and test sets are shown in Figs. 7.16 and 7.17.

Unfortunately, in Fig. 7.16, there is no convergence even using 300 generations. From Fig. 7.16, it can be noticed that the training and testing accuracies of bending to the left can only reach 91.25 and 90.96% (Fig. 7.16b), the training and testing accuracies of bending to the right only reach 94.9 and 94.5% (Fig. 7.16d), while the training and testing accuracies of bending to the left for the 1000 generations can reach 97.79 and 97.65% (Fig. 7.17b), and the training and testing accuracies of bending to the right for the 1000 generations can reach 96.32 and 95.954% (Fig. 7.16d). Although it doesn't look like the results change much in the Loss curves. As can be

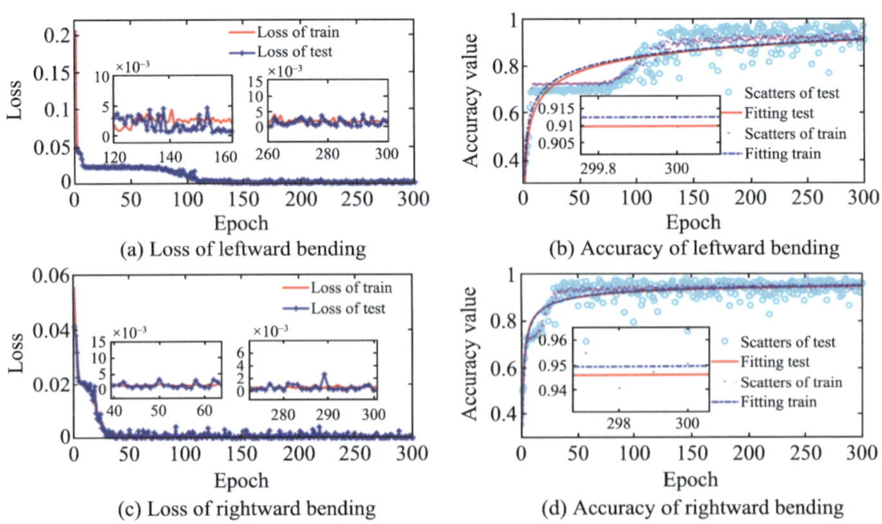

Fig. 7.16 300 iterations loss and accuracy of turning left and right

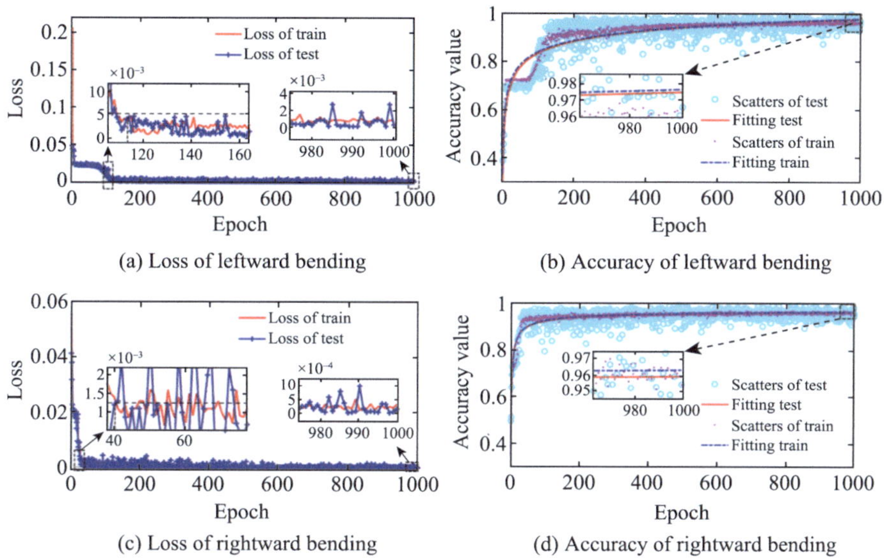

Fig. 7.17 1000 iterations loss and accuracy of turning left and right

seen, the training and testing loss values of the leftward bending motion all reached a stable value of 0.005 at 110 generations (Fig. 7.16a), with final accuracies of 97.79 and 97.65% (Fig. 7.16b), respectively; The training and testing loss functions for the rightward bending motion were around the lower value of 0.001 in 42 generations (Fig. 7.16c), and the final training and testing accuracies are stable at 96.32 and 95.954% (Fig. 7.16d), respectively. They all have a loss of less than 0.0003 for 1000 generations, which again confirms the good performance and the feasibility of the IK-MLP-CR modeling on soft manipulators.

Further, after obtaining the loss and accuracy of the inverse kinematic prediction model (IK-MLP-CR), its grasping performance can be assessed by the grasping error and grasping rates. The pressure sequence (P_2, P_5, P_8) can be input with 1728 sets of 20,000–120,000 Pa with 20,000 Pa intervals. Note that (P_3, P_6, P_9) have the same values as (P_2, P_5, P_8), forming various deformations bent to the left. In the bending schematic to the left of Fig. 7.18, the feature points at the end of bending segments 1~3 can be seen. Since only the inverse kinematics of the plane bending of the robot arm has been modeled, the bending workspace for these points can be converted from the X-Y-Z axis to the X-Z axis in the horizontal plane, as shown in Fig. 7.18a–c. The inverse modeling effect of the root bending segment (Bs No.1) which is near the fixed end of the manipulator is the best, with errors of less than 1 mm in Fig. 7.18c, but the end bending segment (Bs No.3) inverse modeling accumulates the error, with the maximum of 1.53cm and the grasping rate of 97.05% in Fig. 7.18a.

The pressure sequence (P_1, P_4, P_7) can be input with 1728 sets of 20,000–120,000 Pa with 20,000 Pa intervals, forming various deformations for the right bending. The bending workspace in Fig. 7.10, for the feature points at the end of the bending

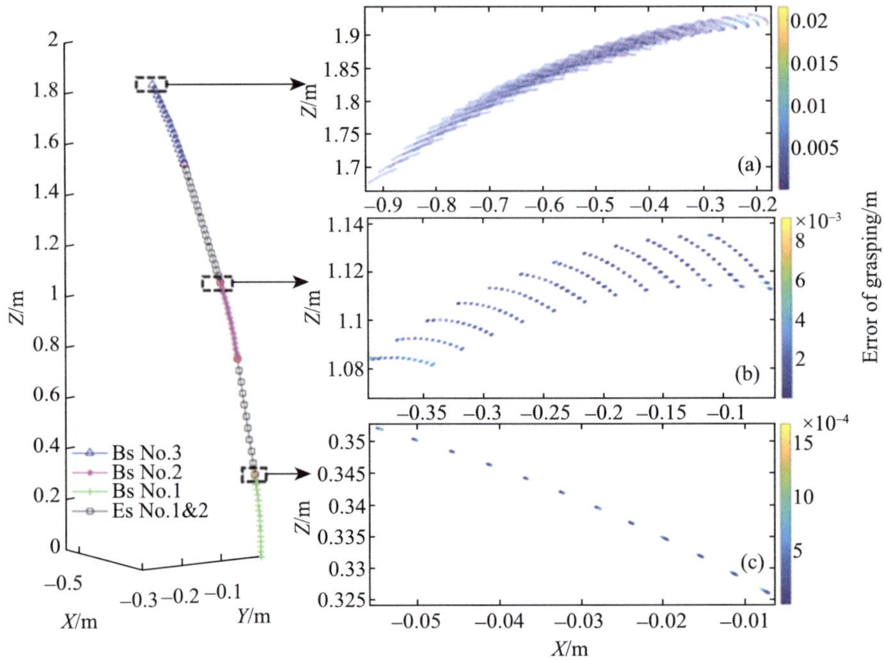

Fig. 7.18 The grasping error of leftward bending: **a** grasping error for the third bending segment end (Bs No. 3); **b** grasping error for the end of the second bending segment (Bs No. 2); **c** grasping error for the end of the first bending segment (Bs No. 1)

segments 1–3, can be converted from the X-Y-Z axis to the X-Z axis in the horizontal plane, i.e. shown by Fig. 7.19a–c. The reachable grasping workspace at the end of the manipulator constitutes a moon-brow shape and additionally indicates the grasping error at the corresponding position. Figure 7.19a shows that the cumulative training grasping error e_1 for the end-effector feature point is relatively large, among these 1728 sample groups, reaching a maximum of 1.13cm. About 28 groups have an error of more than 7mm, reaching a successful grasping rate of 98.38%. From Fig. 7.19c, the max grasping errors of the single segment are only about 0.8mm, achieving a grasping rate of as high as 100%.

The different grasping rates from these two bending can be combined with the information in Fig. 7.12 of Sect. 7.4.1, the triangular profile of bending segments No. 1–No. 3 induces an imbalance between left and right bending on the horizontal plane. This discrepancy elucidates the distinctions observed in Figs. 7.18 and 7.19. On the right side of the soft manipulator (specifically bellows 2, 3, or 5, 6, or 8, 9), there is a lower resultant force in the X-axis, accompanied by a lesser degree of bending in comparison to the right turn. In contrast, a single bellow (bellows 1, 4, or 7) of the soft manipulator yields a greater displacement in the X-axis.

From the above training results, the low loss of under 0.0003 and the successful grasping rates of 97.05 and 98.38% for left and right bends show the feasibility

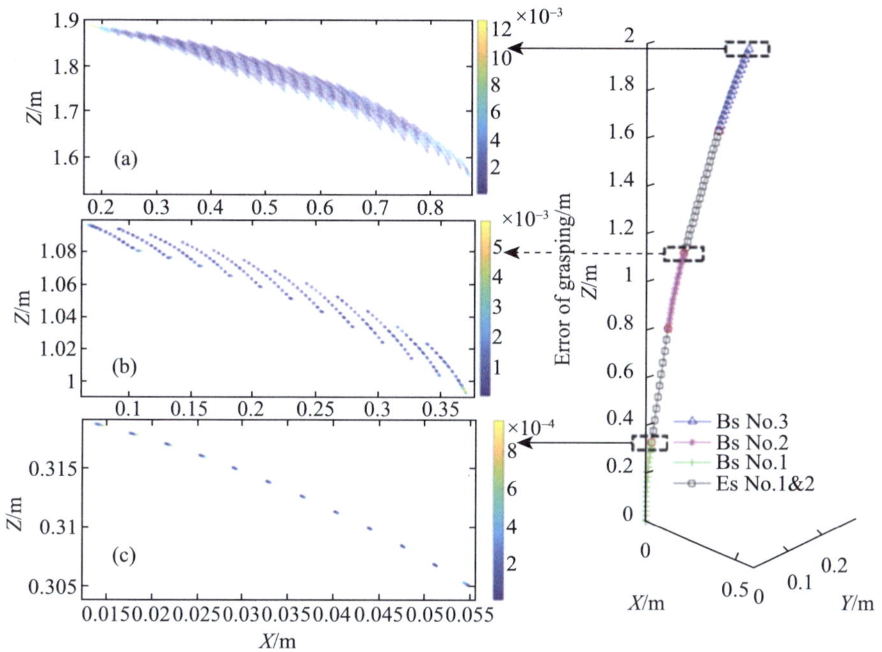

Fig. 7.19 The grasping error of rightward bending: **a** grasping error for the third bending segment end (Bs No. 3); **b** grasping error for the end of the second bending segment (Bs No. 2); **c** grasping error for the end of the first bending segment (Bs No. 1)

of the IK-MLP-CR method and the strong prediction of the unidirectional motion. And obviously, the grasping rate of single-segment bending in Figs. 7.18 and 7.19 achieves as high as 100%. It can be due to the inclusion of the classification model, which greatly reduces the training time and increases the modeling accuracy.

The ground test system of space intelligent soft manipulator consists of three parts: soft manipulator, pneumatic control subsystem and visual localization subsystem. The soft manipulator adopts a modularized design idea, which consists of three bending segments stacked in series, two extended segments, a soft gripper and connecting components; The pneumatic control subsystem includes the air supply, switching valve, pressure relief valve, proportional pressure reducing valve, high-precision pressure sensor, air capacitor, etc.; And the 3D VICON motion tracking system is utilized to measure the position coordinates of multiple feature points by a number of collectors (with an output frequency of 100Hz), a switchboard and a PC mainframe computer, which optically locates the real-time position and attitude information of multiple marker balls distributed on the robotic arm. The whole inflation process and specifications in Table 7.7 are as follows: the air supply provides the air source; The switching valve controls the opening and closing of the pressure transmission; the pressure relief valve REG-1 protects channels 1–9 by determining the release pressure, which is measured in real-time by the high-precision pressure

Table 7.7 Overview of the plant system specifications

System name	Components	Details	Function	Companies
Air-bearing system and platform		Marble, 10 m × 6 m	High precision plane	
Soft manipulator subsystem	Three bending segments	Silicone materials	3D motion	Better silicone rubber
	Two extended segments			
	Connected components			
	A soft gripper	Silicone	Grasping	SRT
Pneumatic control subsystem	Air supply	Under 20 bar	Provide pressure	General
	Switching valve	Ball valve QH-1/2	Turn on and off	FESTO
	Pressure relief valve	QS-6	Adjustment and saving	
	Air-capacitor	VZS-20-B	Compensation of pressure fluctuations	
	Proportional pressure reducing valve	VEMD	Flow control	
	Pressure sensor	SDE5	Pressure measuring	
Visual positioning system	3D VICON motion tracking system (precision 1 mm) / Several collectors	Motion capture cameras		VICON
	Several collectors	Reflective marker balls		
	A switch			
	a PC	Tracker/Nexus software	Movement information	Lenovo

sensor PT-1. Since this work does not analyze the robotic hand in detail, and the preset value of the input of the extension section is set to a constant value to keep the length of the extension section the same, this work only needs to consider the pneumatic control of the 9 channels in the bending section.

Taking a bending process of the soft manipulator, the corresponding time series motion of its coordinates, velocity and acceleration in X or Z coordinates is plotted as shown in Fig. 7.20. The whole process of pressurization, inflation, bending, being stable, back to the original state and being original takes about 10–20s, which can be seen as a one-way quasi-static process with about 0.17 m/s. The slow movement of the flexible arm is mainly due to the swinging of the long arm and the effect of the inflatable effect. The mentioned bending process of the soft manipulator is categorized into: ① before bending (I); ② bending (I-V); ③ being the stable state (VI); ④ back to the original state (VII-XI); ⑤ being the original state (XI-XIV). It can be seen that the pressurized reaction time ($0 \sim t_1$) is about 3s, the bending time ($t_1 \sim t_5$) during the imbalance of the horizontal force is about 4s, and it has come to be static ($t_5 \sim t_7$) for about 3s after bending into a certain shape. After that, the manipulator is deflated and reverts back to the original position ($t_7 \sim t_{11}$) in only 2s, which can be seen as only half of the time of the process (I~V). And the state of (XI~XIII) is the original state, which also needs to offset the buffer and then reach a standstill. Besides, in Fig. 7.20b, the velocity in the X-axis direction is a process of acceleration and then roughly at a uniform velocity from $t_2 \sim t_5$, with an

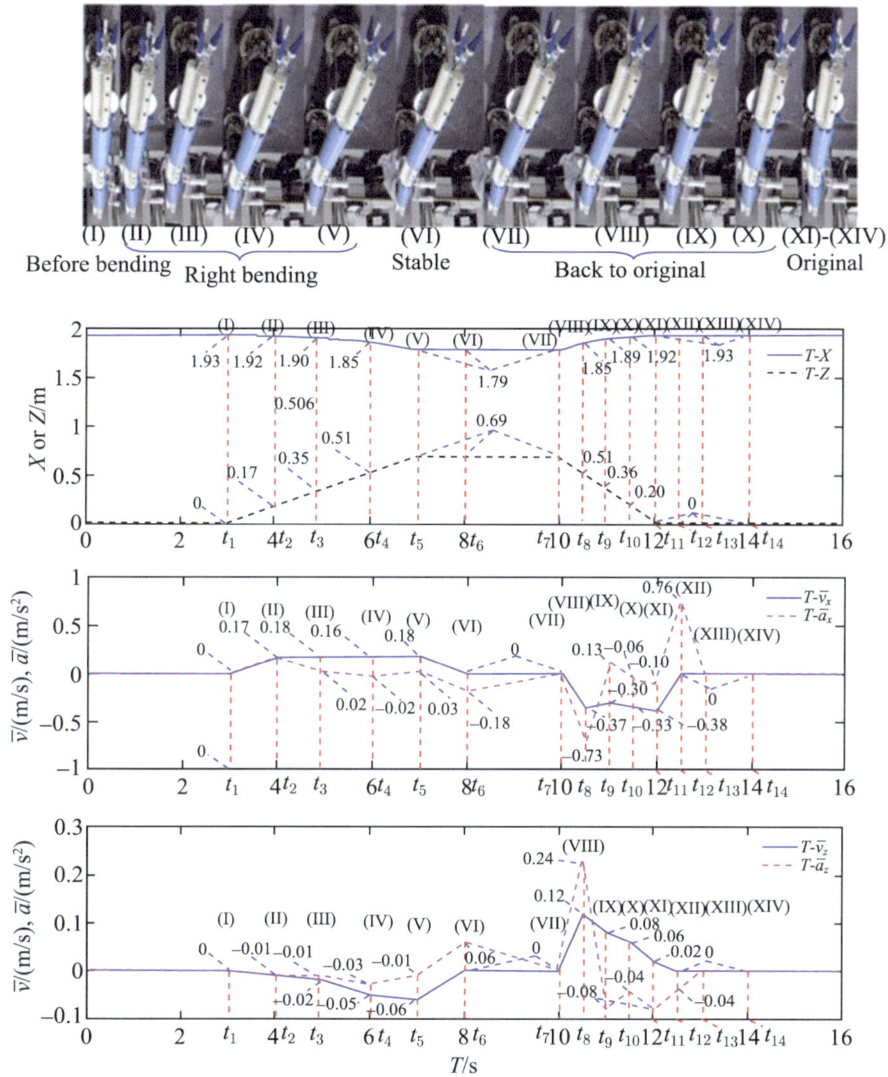

Fig. 7.20 Experimental time series schematic of X and Y coordinates for a large degree of bending

average velocity of about 0.17 m/s; In the Z direction of Fig. 7.20c, there is about a uniform accelerated motion with an acceleration of 0.01 m/s^2, but at the time of t_4, there is an accelerated motion of 0.03 m/s^2. This phenomenon occurs due to the constraints of the extended segment as an elongated structure, which requires more internal pressure to break free.

Contact motions can be divided into two types: one for any position of the body and the other for the end of the manipulator. The former involves contact mechanics,

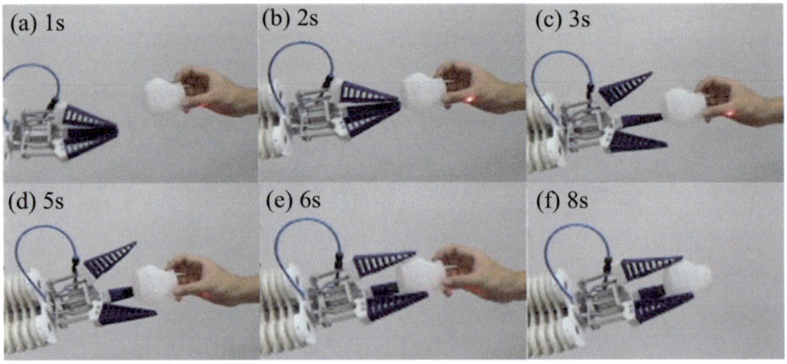

Fig. 7.21 Experimental time schematic of end-contact forces with grasping a small 3D-printed white bottle

while the latter is mainly concerned with grasping irregular objects as well as end-contact forces. The study of grasping a small 3D-printed white bottle is given in Fig. 7.21. The robot has been tested to grasp objects with a maximum weight of 1 kg and surface areas ranging from 1 cm $\times$ 1 cm to 10 cm$\times$10 cm.

7.5 Summary

In this study, the design, modelling and implementation of a foldable pneumatic soft manipulator are presented. The design of the soft manipulator exhibits a novel performance of the transformation between the contracted state and the expanded state, which presents broad applications in a number of fields, especially in space manipulation tasks. When it is in the contracted state, the soft manipulator can conserve storing volume; When it switches to the expanded state, the soft manipulator can conduct the operating tasks in a wide workspace. Based on this design, a kinematic model of the soft manipulator is proposed based on the piecewise constant-curvature assumption.

In the numerical calculation and analysis, two different control strategies are designed to achieve different shape deformations of the soft manipulator. All the numerical calculation results indicate that no disturbance occurs in the numerical results of the kinematic model, which are consistent with the physical laws of the soft manipulator. From the workspace analysis, it is clear that the soft manipulator has a much wider workspace when it is in the expanded state.

To validate the prediction accuracy of the kinematic model of the soft manipulator without considering the gravity effect, an experimental setup using air-floating devices based on a marble platform is designed and developed. The numerical results of the kinematic model are also converted to the plane condition to compare with the experimental results. By comparing the deformation patterns and the bending trends, the model prediction and the experimental results agree well with each other, thereby validating the correctness of the kinematic model. However, differences between the model prediction and the experiment still exist. To reduce these differences, both the kinematic model and the experimental system should be improved. For kinematic models, geometrically accurate models and other modelling methods with more accurate measurement of the internal friction should be considered; While for the experiment system, embedded shape detection methods, such as FBGS, should be considered.

Finally, based on the kinematic model, a method combining classification and regression based on multi-layer perceptron for inverse kinematic modelling, namely the IK-MLP-CR method is proposed. It can classify the bending states and predict the whole manipulator shape and greatly reduce the amount of training data and training time. In general, the investigation in this chapter provides a meaningful attempt for the foldable soft manipulator applied in future space or ground missions.

References

1. Trivedi D, Lotfi A, Rahn CD (2008) Geometrically exact models for soft robotic manipulators. IEEE Trans Rob 24(4):773–780
2. Webster RJ, Jones BA (2010) Design and kinematic modeling of constant curvature continuum robots: a review. Int J Robot Res 29(13):1661–1683
3. Aikenhead BA, Daniell RG, Davis FM (1983) Canadarm and the space shuttle. J Vac Sci Technol Vac Surf Films 1(2):126–132
4. Grazioso S, Di Gironimo G, Siciliano B (2019) A geometrically exact model for soft continuum robots: the finite element deformation space formulation. Soft Robot 6(6):790–811
5. Renda F, Armanini C, Lebastard V et al (2020) A geometric variable-strain approach for static modeling of soft manipulators with tendon and fluidic actuation. IEEE Robot Autom Lett 5(3):4006–4013
6. Chen Q, Yang L, Zhao Y, et al (2024) Shortest path in LEO satellite constellation networks: an explicit analytic approach. IEEE J Select Areas Commun 1–13
7. Chen Q, Giambene G, Yang L et al (2021) Analysis of inter-satellite link paths for LEO mega-constellation networks. IEEE Trans Veh Technol 70(3):2743–2755
8. Li J, Chen X, Su Y et al (2022) Kinematic analysis of soft continuum manipulators based on sparse workspace mapping. IEEE Robot Autom Lett 7(2):5055–5062
9. Amanov E, Nguyen TD, Burgner-Kahrs J (2021) Tendon-driven continuum robots with extensible sections—a model-based evaluation of Path-following Motions. Int J Robot Res 40(1):7–23
10. Wang Y, Wu Z, Wang L et al (2022) Inverse kinematics and dexterous workspace formulation for 2-Segment continuum robots with inextensible segments. IEEE Robot Autom Lett 7(1):510–517
11. Wu YY, Tan N (2020) Model-less feedback control for soft manipulators with Jacobian adaptation. In: International symposium on autonomous systems (ISAS). Guangzhou, China, pp 217–222
12. Coevoet E, Escande A, Duriez C (2017) Optimization-based inverse model of soft robots with contact handling. IEEE Robot Autom Lett 2(3):1413–1419
13. Du T, Hughes J, Wah S et al (2021) Underwater soft robot modeling and control with differentiable simulation. IEEE Robot Autom Lett 6(3):4994–5001
14. Li X, Zheng T, Sui D et al (2022) A 3D printed variable cross-section pneumatic soft manipulator with high-precision positioning capability: design and control implementation. Sens Actuat A 342:113644

15. Zou S, Lyu Y, Qi J et al. A deep neural network approach for accurate 3D shape estimation of soft manipulator with vision correction. Sens Actuat A Phys 344:113692
16. Liu Z, Zhang X, Wang J et al. Design, modelling and implementation of a foldable pneumatic soft manipulator. Micrograv Sci Technol 34(3):44
17. Haykin S (2009) Neural networks and learning machines Pearson. Education India
18. Giorelli M, Renda F, Calisti M et al (2015) Neural network and Jacobian method for solving the inverse statics of a cable-driven soft arm with nonconstant curvature. IEEE Trans Robot 31(4):823–834

Chapter 8
Design and Experimental Validation of a Cable-Driven Continuum Manipulator and Soft Gripper

The design, fabrication and experiment are still the main concerns for the soft manipulator. The modeling and analysis are largely different according to different designs. In this chapter, a cable-driven continuum manipulator and its soft gripper are designed and developed. The theoretical model of the continuum manipulator is proposed and the motion performance is analyzed. The topological optimization method is introduced into the design process of the soft gripper. In Sect. 8.1, the general design of the continuum manipulator system, the design of the continuum arm, and the topological optimization of the soft gripper are introduced. In Sect. 8.2, the experiment system of the continuum manipulator is developed. The capture test of the soft gripper and the motion control of the continuum manipulator are conducted to validate the performance of the experiment system. Finally, the conclusions are presented.

8.1 Design of a Cable-Driven Continuum Manipulator and Soft Gripper

8.1.1 General Design Schematic

The soft robot system in this section is made up of a continuum arm, a soft gripper and a cable-driven control system. The continuum arm is driven by a series of cables and is able to bend flexibly. The soft gripper is also driven by the cable and can adapt to the shape of the objects. The cable-driven control system includes a numbers of motors and pulleys, which are all mounted on the frame and the base. The general design schematic is shown in Fig. 8.1.

The continuum manipulator is made up of two sections, including nine threading plates and eight spring steels, as shown in Fig. 8.2. Section I is from the top to the end and section II is from the middle to the end. The continuum manipulator is driven by eight cables. Thus, the continuum manipulator has eight control freedoms. The four cables from the top to the end drive the section I to bend and deform. The other

X. Zhang et al., *Design and Implementation of Soft Robotic Manipulators*,
https://doi.org/10.1007/978-981-95-1657-5_8

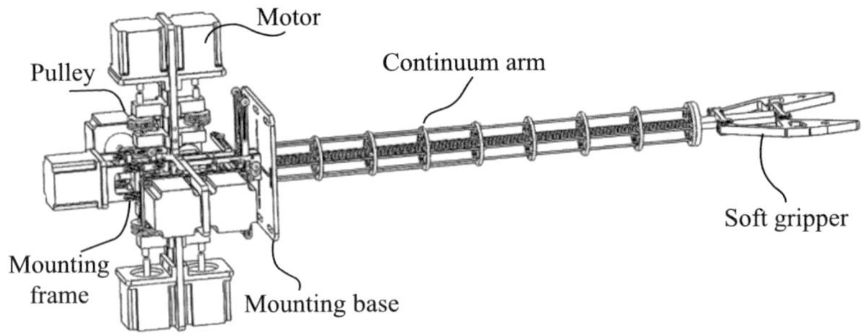

Fig. 8.1 General design schematic

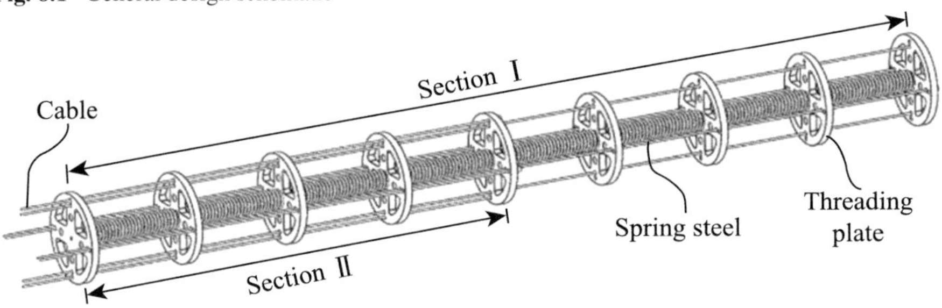

Fig. 8.2 Design of cable-driven continuum manipulator

four cables from the middle to the end drive the section II to bend and deform. The coupling effect of the two sections can make the continuum manipulator bend as the "S" shape. The threading plate is made of aluminum. There are eight threading holes for the eight cables, and features a mounting site for connecting the springs.

8.1.2 Topological Optimization Design of a Soft Gripper

It is known that human hands have the ability to grasp an object with enough force and wrapping area [1, 2]. The force and wrapping area are two important factors to determine the grasping performance [3]. Thus, the output force and the output displacement are considered. In order to conduct the topological optimization design [4, 5], two technical indexes, mechanical advantage (MA) and geometry advantage (GA), are defined. MA is the ratio of output force and input force, GA is the ratio of output displacement and input displacement. MA determines the grasping force while GA determines the wrapping area. The output force can be described using mutual strain energy (MSE) [6], and the output force can be described using structure softness (SE_o).

$$\mathrm{MSE} = \boldsymbol{U}_{\mathrm{o}}^{\mathrm{T}} \boldsymbol{K} \boldsymbol{U}_{\mathrm{i}} \tag{8.1}$$

$$\mathrm{SE}_{\mathrm{o}} = \frac{1}{2} \boldsymbol{U}_{\mathrm{o}}^{\mathrm{T}} \boldsymbol{K} \boldsymbol{U}_{\mathrm{o}} \tag{8.2}$$

where $\boldsymbol{U}_{\mathrm{i}}$ is the displacement field produced by the input load; $\boldsymbol{U}_{\mathrm{o}}$ is the displacement field produced by the output virtual load; $\boldsymbol{K}$ is the parameter transformation matrix. The optimization objective function is presented as follows:

$$-\frac{\mathrm{MSE}}{\omega \mathrm{SE}_{\mathrm{i}} + (1-\omega)\,\mathrm{SE}_{\mathrm{o}}} \tag{8.3}$$

where

$$\mathrm{SE}_{\mathrm{i}} = \frac{1}{2}\boldsymbol{U}_{\mathrm{i}}^{\mathrm{T}}\boldsymbol{K}\boldsymbol{U}_{\mathrm{i}} \tag{8.4}$$

In order to describe the wrapping area more accurately, the divisional multi input and output method is introduced. The core concept is to divide the design area into numbers of small areas, in which the single input or output point is selected. Thus, the optimization objective function can be improved as follows:

$$-\frac{(1-\xi)\left(\frac{1}{mn}\sum_{i=1}^{n}\sum_{j=1}^{m} S_{\mathrm{mut1},ij}\right)+\xi\left(\frac{1}{pn}\sum_{i=1}^{n}\sum_{k=1}^{p} S_{\mathrm{mut2},ik}\right)}{2\eta\left(\frac{1}{n}\sum_{i=1}^{n} S_{\mathrm{in},i}\right)+2(1-\eta)\left[(1-\xi)\left(\frac{1}{m}\sum_{j=1}^{m} S_{\mathrm{out1},j}\right)+\xi\left(\frac{1}{p}\sum_{k=1}^{p} S_{\mathrm{out2},k}\right)\right]} \tag{8.5}$$

where η is used to control the relationship between the mutual strain energy and the structure strain energy, ζ is used to control the weight relationship between two output areas. The number of input points is n and the number of output points in two output areas are separately m and p. The topological optimization design problem in this investigation can be described as follows:

$$\begin{cases} \min: -\dfrac{(1-\xi)\left(\frac{1}{mn}\sum_{i=1}^{n}\sum_{j=1}^{m} S_{\mathrm{mut1},ij}\right)+\xi\left(\frac{1}{pn}\sum_{i=1}^{n}\sum_{k=1}^{p} S_{\mathrm{mut2},ik}\right)}{2\eta\left(\frac{1}{n}\sum_{i=1}^{n} S_{\mathrm{in},i}\right)+2(1-\eta)\left[(1-\xi)\left(\frac{1}{m}\sum_{j=1}^{m} S_{\mathrm{out1},j}\right)+\xi\left(\frac{1}{p}\sum_{k=1}^{p} S_{\mathrm{out2},k}\right)\right]} \\ \text{subject to}: \boldsymbol{K}\boldsymbol{U}_1 = F_1,\ \boldsymbol{K}\boldsymbol{U}_2 = F_2,\ \boldsymbol{K}\boldsymbol{U}_3 = F_3 \\ \qquad \sum_{i=1}^{N} x_i V^e = V_f \\ 0 < x_{\min} < x_i < 1 \end{cases} \tag{8.6}$$

The topological optimization design process for the soft gripper is shown in the following Fig. 8.3.

In the optimization process, the Zig-Zag phenomenon, which directly affects the three-dimensional design of the structure, appears at the edge of the structure, as

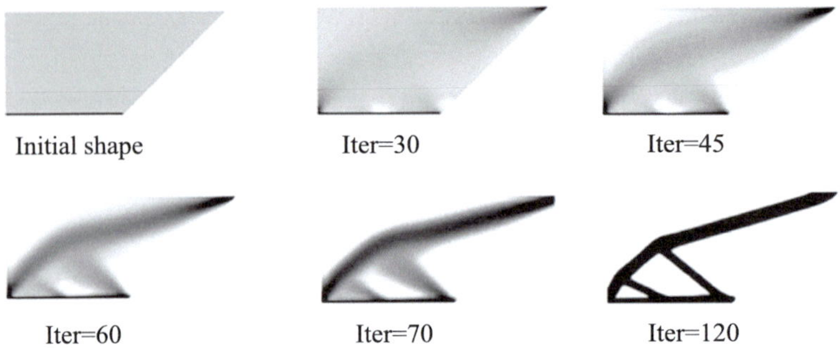

Fig. 8.3 Iterative process of topology optimization

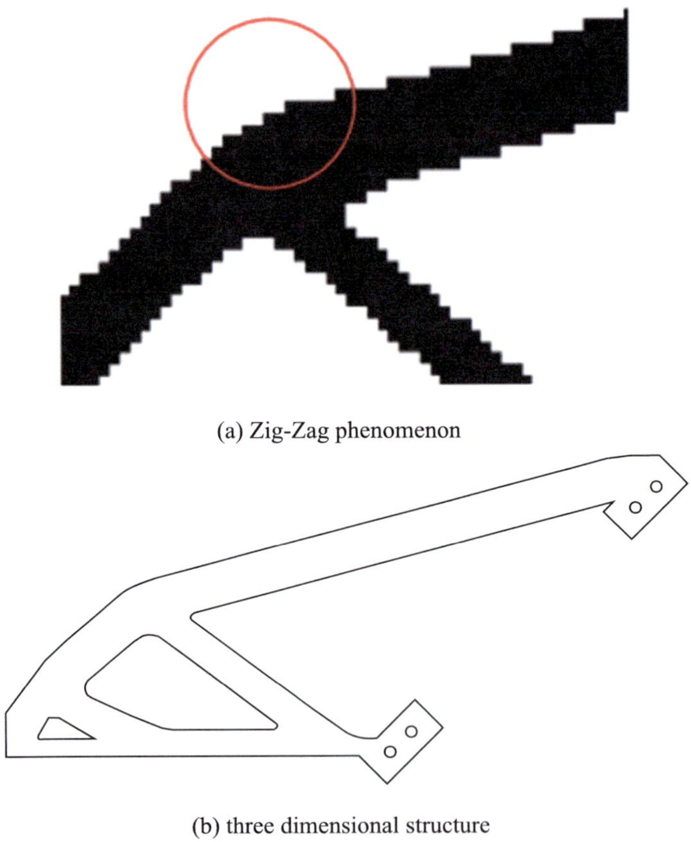

Fig. 8.4 Post-processing results

shown in Fig. 8.4a. In order to obtain a smooth edge, the level-set method is adopted in the post-processing of the designed structure. Finally, the three dimensional structure of the soft gripper is obtained, shown in Fig. 8.4b.

8.2 Experimental Validation

8.2.1 Experiment System

The experiment system of the continuum manipulator includes the continuum arm, the soft gripper, and the cable-driven control system, as is shown in Fig. 8.5a. The continuum arm and the cable-driven control system are mounted on the aluminum

Fig. 8.5 Experiment system

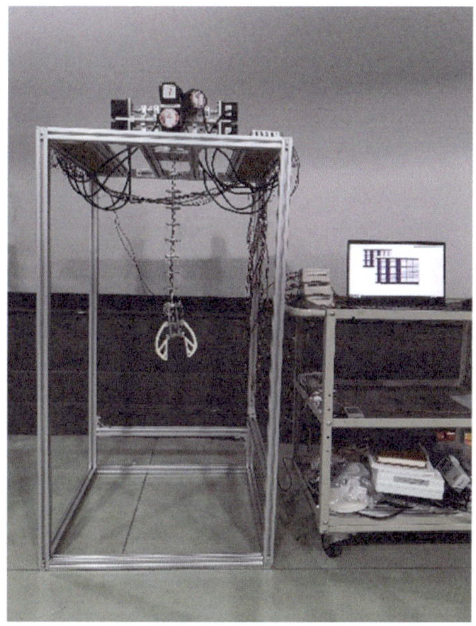

(a) Prototype and experiment system

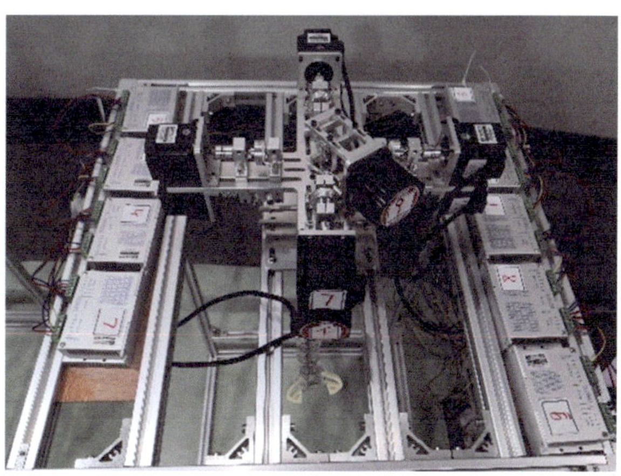

(b) Control system

frame with a suspension type. The cable-driven control system is shown in Fig. 8.5b. The diameter of the continuum manipulator is 50 mm. The diameter of the spring steel is 15 mm, and the length is 50 mm. The total length of the continuum manipulator is 450 mm. The cable is made of steel and its diameter is 0.5 mm. The maximum tension can reach 30 kg. The soft gripper is mounted at the end of the continuum arm. It is made using 3D printing technology with TPU [7]. The shore hardness of the TPU is D60. The capture test of the soft gripper and the motion control of the continuum manipulator will be performed in the following investigation.

8.2.2 Capture Test of the Soft Gripper

The soft gripper is composed of two fingers, as is shown in Fig. 8.6. The grasp contact area of each finger is about 70 mm × 10 mm. The actuation displacement is about 35 mm. When the soft gripper grasps objects, it can be seen that the fingers deform adaptively according to the shape of the objects, such as the water bottle, the egg, the table tennis, the motor, et al. As the weight of the objects increases, it can be seen that the adaptive deformation of the soft fingers increases. It is because the large structure deformation means the large force and wrapping area. In the following test, the weight from 3 g (a table tennis) to 500 g (a motor) is tested, shown in Fig. 8.6.

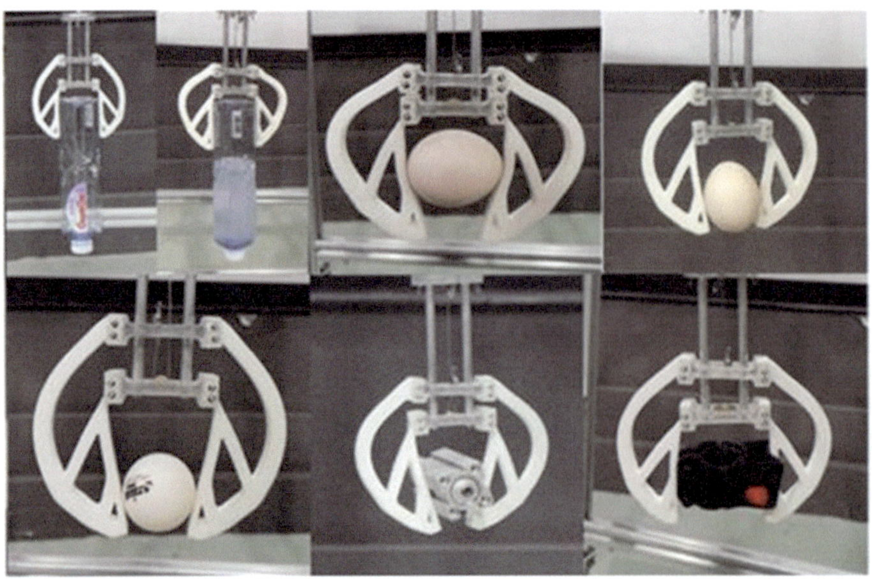

Fig. 8.6 Capture test of the soft gripper

8.2.3 Motion Control of the Continuum Manipulator

The continuum manipulator is driven using nine cables, in which eight cables are used to control the continuum arm and one cable is used to control the soft gripper. In Fig. 8.7a, the motion control without payload is tested. It can be seen that the continuum manipulator can move according to different control patterns. In Fig. 8.7b, the motion control with a payload of 200 g is tested. It can be seen that the continuum manipulator has the ability to transport the object to different sites. Currently, it is just validated that the continuum manipulator as well as the soft gripper are able to be controlled by the cable-driven control system. The deep research on the accurately modeling and control will be studied in our future work.

(a) Motion control without payload

(b) Motion control with payload (200g)

Fig. 8.7 Motion control of the continuum manipulator

8.3 Summary

In this investigation, a cable-driven continuum manipulator as well as its soft gripper were designed and developed. The cable-driven mode can provide a large control moment as well as good motion flexibility. The coupling effect of the two actuation sections can make a complex movement pattern. The soft gripper is designed by using the topological optimization method, and its final design structure is far different from the traditional result. It is validated that the soft gripper has a good grasping performance from the capture test results. The design and experiment of the soft gripper show us a wide prospect of the topological optimization method in the research of the soft robot. Currently, the continuum manipulator has only the ability to move according to the different pulling patterns of the cables. It is a big challenge for the current manipulator to keep stable during the motion control process. Next, the structure of the proposed continuum manipulator needs to get improved. The accurate model-driven control algorithm based on the kinematical and dynamical model of the cable-driven continuum manipulator will be studied in the future work.

References

1. Feix T, Romero J, Schmiedmayer HB et al (2015) The grasp taxonomy of human grasp types. IEEE Trans Human Mach Syst 46(1):66–77
2. Choi C, Schwarting W, DelPreto J et al (2018) Learning object grasping for soft robot hands. IEEE Robot Autom Lett 3(3):2370–2377
3. Della Santina C, Arapi V, Averta G (2019) Learning from humans how to grasp: a data-driven architecture for autonomous grasping with anthropomorphic soft hands. IEEE Robot Autom Lett 4(2):1533–1540
4. Eschenauer HA, Olhoff N (2001) Topology optimization of continuum structures: a review. Appl Mech Rev 54(4):331–390
5. Wang R, Zhang X, Zhu B et al (2020) Topology optimization of a cable-driven soft robotic gripper. Struct Multidiscip Optim 62:2749–2763
6. Taylor JE, Bendsoe MP (2001) A mutual energy formulation for optimal structural design. Struct Multidiscip Optim 22:95–101
7. Chen Q, Cao PF, Advincula RC (2018) Mechanically robust, ultraelastic hierarchical foam with tunable properties Via 3D printing. Adv Func Mater 28(21):1800631

MIX
Papier aus verantwortungsvollen Quellen
Paper from responsible sources
FSC® C105338

If you have any concerns about our products,
you can contact us on
ProductSafety@springernature.com

In case Publisher is established outside the EU,
the EU authorized representative is:
Springer Nature Customer Service Center GmbH
Europaplatz 3, 69115 Heidelberg, Germany

Printed by Libri Plureos GmbH
in Hamburg, Germany